United States
Department of
Agriculture

Foreign
Agricultural
Service

Circular Series
WAP 8-14
August 2014

# World Agricultural Production

I0484313

**Canada Rapeseed:  Area Down from Late Planting and Floods**

USDA estimates 2014/15 Canada rapeseed production at 15.25 million tons, down 2.9 percent from last month and down 15 percent from last year's record level. Harvested area is estimated at 7.7 million hectares, down 3.9 percent from last year and down 3.8 percent from last month. The reduction in estimated area is due to unseeded acreage and late June flooding that impacted southeast Saskatchewan and southwest Manitoba. Yield is forecast at 1.98 tons per hectare, down 12 percent from last year's record but up 0.9 percent from last month.

Rapeseed development in the Western Prairies is still behind normal because of wet and cool weather conditions coupled with late planting. However, warmer temperatures in late July and sufficient precipitation have advanced crop maturation. The majority of the crop is setting pods. Analysis of satellite-derived vegetation indices show that crop development across most of the western prairies is slightly above the 5-year average. However, crop vigor in southeast Saskatchewan and southwestern Manitoba, which account for approximately 5 percent of total rapeseed production, is likely to be slightly below the short-term mean. Recent provincial reports from the major producers, Saskatchewan and Alberta, have rated the rapeseed crop as 70 percent good to excellent. (*For more information, please contact Arnella Trent at 202-720-0881.*)

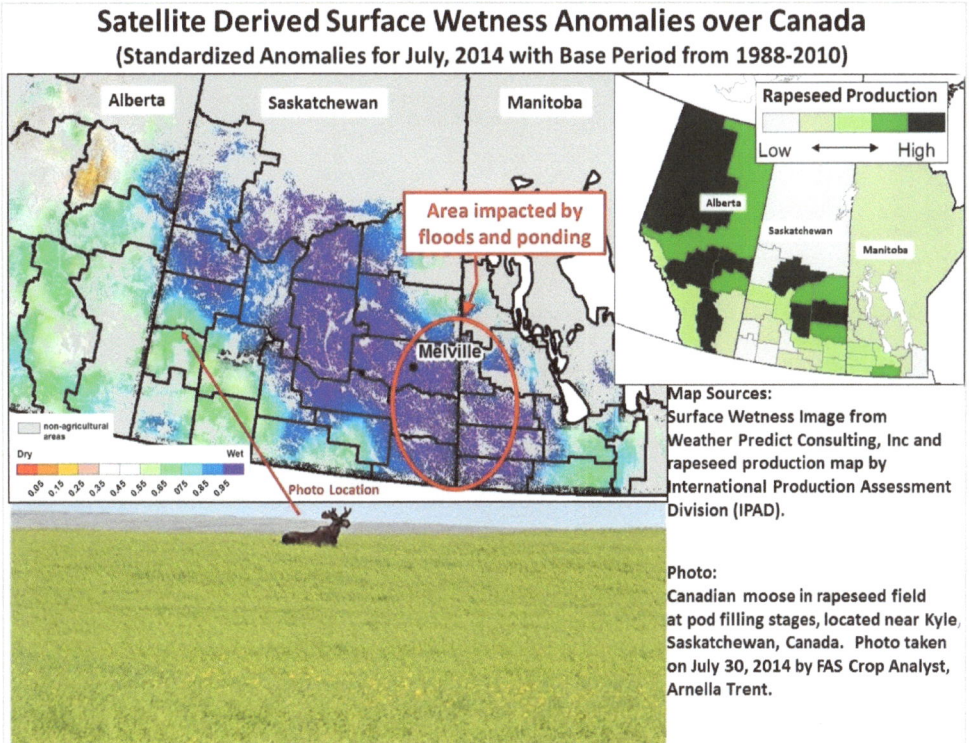

**Satellite Derived Surface Wetness Anomalies over Canada**
(Standardized Anomalies for July, 2014 with Base Period from 1988-2010)

Alberta   Saskatchewan   Manitoba

Area impacted by floods and ponding

Melville

non-agricultural areas

Dry       Wet

0.05  0.15  0.25  0.35  0.45  0.55  0.65  0.75  0.85  0.95

Photo Location

Rapeseed Production

Low ← → High

Alberta
Saskatchewan
Manitoba

Map Sources:
Surface Wetness Image from Weather Predict Consulting, Inc and rapeseed production map by International Production Assessment Division (IPAD).

Photo:
Canadian moose in rapeseed field at pod filling stages, located near Kyle, Saskatchewan, Canada.  Photo taken on July 30, 2014 by FAS Crop Analyst, Arnella Trent.

**Approved by the World Agricultural Outlook Board**

**Russia Wheat:  Outstanding Winter-Wheat Yields Boost Estimated Output**

USDA estimates Russia wheat production at 59.0 million tons, up 6.0 million or 11 percent from last month and up 6.9 million or 13 percent from last year.  Harvested area is estimated at 23.8 million hectares against 23.4 million last year.  Yield is estimated at a record 2.48 tons per hectare, up 11 percent from last month, up 12 percent from last year, and 18 percent above the 5-year average.  The month-to-month increase is based on harvest reports from the Ministry of Agriculture indicating record wheat yields in European Russia, where nearly all the country's winter wheat is grown and where harvest is nearing completion.  Winter wheat typically comprises about two-thirds of Russia's total wheat output.

Analysis of weather data and satellite-derived vegetation indices show that conditions for spring wheat are better than last year in the Ural and Volga Districts (which account for approximately 20 and 30 percent of Russia's spring-wheat output, respectively), but slightly worse than last year in the Siberian District, which produces nearly half of the country's spring wheat.  Spring-wheat harvest will begin in late August and continue until the end of October. (*For more information, please contact Mark Lindeman at 202-690-0143.*)

### Russia Spring-Wheat Zone:  MODIS NDVI Anomaly
### July 12-19, 2014

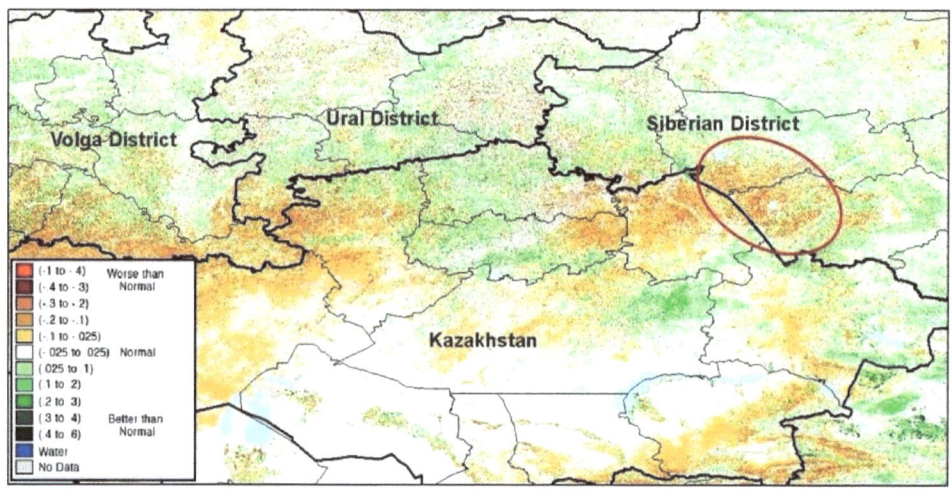

Vegetative indices indicate that wheat conditions were generally above average in the Ural and Volga Districts in mid-July as the crop was approaching the flowering stage. In key parts of the Siberian District (circled), conditions were below normal due in part to dryness during June and early July.

Source:  NASA-GSFC-GIMMS/USDA-FAS-OGA

**China Wheat:  Record Yield and Production**

USDA estimates China's 2014/15 wheat crop at a record 126.0 million tons, up 2.0 million or 1.6 percent from last month and up 4.1 million or 3.3 percent from last year.  Area is estimated at 24.1 million hectares, down slightly from last month and last year.  The estimated yield of 5.23 tons per hectare is up 2.0 percent from last month and up 3.4 percent from last year's record yield.

Winter wheat accounts for about 95 percent of China's total wheat output. China's 2014/15 winter wheat crop was planted in October 2013 and harvested in June 2014. Overall, the weather during the growing season was favorable. Temperatures and soil moisture were suitable at planting, and the crop entered dormancy in good condition. The weather in December and January was seasonably dry, and temperatures were relatively mild. Drought conditions developed in a few wheat areas, which increased the need for supplemental irrigation. However, the arrival of cool, wet weather in February raised soil moisture levels as the crop came out of dormancy, and timely rainfall in April and May created favorable conditions for reproduction and grain fill. Insect and disease problems were well controlled this year. The crop matured ahead of schedule, and mostly dry weather in June allowed the harvest to be completed with few delays or post-harvest losses. According to the Ministry of Agriculture, wheat yield increased in each of China's 11 main wheat growing provinces, and the quality of the crop was better than average.

China's small spring wheat crop (less than 5 percent of total production) is grown in the northern and western parts of the country. Harvesting will be complete in August. *(For more information, please contact Paulette Sandene at 202-690-0133.)*

**Russia Sunseed: Year-to-Year Reduction in Sown Area**

USDA estimates Russia sunseed production at 9.8 million tons, down 0.4 million or 3.9 percent from last month and down 0.8 million or 7.1 percent from last year. The month-to-month decrease is based on preliminary sown-area data released by Rosstat, the government statistical agency. Planted area was reported by Rosstat at 6.8 million hectares, against 7.3 million last year. USDA forecasts harvested area at 6.5 million hectares, down 0.3 million from last year. Conditions have been generally favorable for sunflowers in most key growing regions, and yield will also benefit from a 15-percent increase in the import of hybrid sunflower seed. Hybrid-seed imports have doubled in only four years, and the increased use of hybrid seed and other technical improvements have fueled a 50-percent increase in yield between 2004 and 2013. The forecast yield for 2014/15 is essentially unchanged at 1.51 tons per hectare, down 2.9 percent from last year but 20 percent above the 5-year average. Harvest will begin in September. *(For more information, please contact Mark Lindeman at 202-690-0143.)*

**EU Corn: Production Increases in Response to Heavy July Rain**

USDA estimates the 2014/15 European Union (EU) corn crop at 67.0 million tons, up 1.4 million or 2.1 percent from last month and up 3.1 million or 4.8 percent from last year. Harvested area is estimated at 9.6 million hectares, up 0.1 million from last month but down 0.2 million or 2.4 percent from last year's harvested area. Yield is estimated at 7.02 tons per hectare, above last month's 6.88 tons, and last year's 6.54 tons. The five-year average yield is 6.77 tons per hectare.

Heavy rainfall during the month of July beneficially increased soil moisture levels throughout Europe for summer crops. While the precipitation was optimally timed for corn development, the negative aspect was that it was detrimental to wheat quality and has made wheat harvesting difficult.

Drought conditions existed during winter and spring but ended in June when successive storms brought much needed relief to the corn crop. In Europe, corn typically pollinates in July, and then proceeds into grainfill during August. High soil moisture levels during these critical development periods should substantially bolster yields. Often, much of the EU's corn belt is lacking soil moisture during the hot and dry late summer months, but not in 2014. Satellite-derived vegetation indices graphically depict improved vegetation vigor and higher potential yield during the peak summer crop growth.

The largest individual EU monthly corn change was for Romania, up 0.45 million tons to 9.8 million tons. Other monthly increases include Hungary (up 0.2 million), Bulgaria (0.2 million), and Italy (0.2 million). (*For more information, please contact Bryan Purcell at 202-690-0138.*)

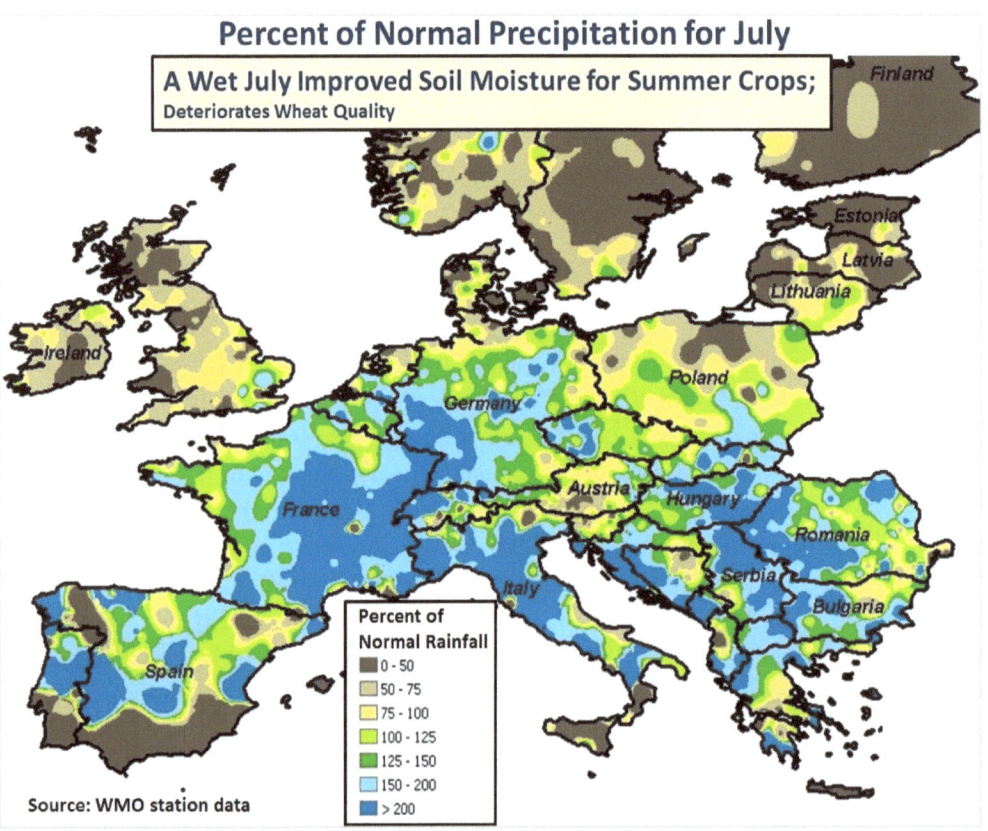

**India Corn: Area Declines from Dry Weather during Planting Season**

India's 2014/15 corn production is forecast at 21 million tons, down 4.1 percent from last month and down 3.19 million or 13 percent from last year. Area is forecast at 8.6 million hectares, down 4.4 percent from last month and down 9.5 percent from last year. Yield is forecast at 2.44 tons per hectare, unchanged from last month, down 4.3 percent from last year, and 1.2 percent higher than the five-year-average.

Roughly 80 percent of India's corn is produced during the *kharif* or monsoon season (June-October) and the rest is grown during the *rabi* season (September-April). According to the Government of India's Meteorological Department, the advance of the 2014 southwest monsoon into the major cotton growing areas of north central and northwestern India improved during the month of July from 35 percent to approximately 25 percent below normal. This year's monsoon was delayed by more than two weeks. At the end of July, regional monsoon rainfall deficits are reported at 49 percent of normal in northwest India, 47 percent in central India, 26 percent in the Southern Peninsula, and 22 percent in northeast India. Generally, the Indian monsoon is characterized as normal at 95 to 105 percent of the long-term-average (LTA), below normal at 90 to 95 percent, and drought at less than 90 percent.

Across the majority of corn growing states (Rajasthan, Karnataka, Utter Pradesh, Madhya Pradesh, Andhra Pradesh, Bihar, and Gujarat), the season started as one of the driest resulting in significant planting delays due to the late start and poor distribution of the rainfall. Estimated area is substantially decreased this month as the window of planting opportunity closes. (*For more information, please contact Dath Mita at 202-720-7339.*)

**India Rice: Production Prospects Continue to Decline from Reduced Area**

India 2014/15 rice production is forecast at 103.0 million tons, down 1.0 percent from last month and down 4.94 million or 3.1 percent from last year's record. Area is forecast at 43.4 million hectares which is 0.54 million less than last year. Yield is forecast at 3.56 tons per hectare, down 1.9 percent from last year.

Rice transplanting is in progress although there were delays in some regions due to the late start, erratic beginning, and poor distribution of the monsoon rainfall. The recent boost in rainfall over eastern India (Bihar, Utter Pradesh, Jharkhand, Andhra Pradesh, Orissa and West Bengal) promoted a better environment for transplanting and early crop development. Overall the majority of key rice growing areas in the eastern states have experienced below normal to approximately normal rainfall. Above-average rainfall is needed during the next few weeks to promote further transplanting and improve the current crop yield expectations.

The southwest monsoon accounts for 70 percent of annual rainfall in India. This year's monsoon was delayed by more than two weeks.

India's rice is produced throughout the calendar year starting with *kharif* rice that accounts for 85 percent of total rice production and is grown in March – February, followed by *rabi* rice accounting for 15 percent and grown in November-June. Almost 50 percent of *kharif* rice is irrigated and mainly produced in the northwest (Punjab, Haryana, Utter Pradesh), northeast and in the Southern Peninsula. *Rabi* rice is 100 percent irrigated and is produced mainly in the east and northeast states (West Bengal, Andhra Pradesh, Orissa, Assam, and Tamil Nadu). The majority of irrigation is by the use of canals, tube wells and diesel pumps. (*For more information, please contact Dath Mita at 202-720-7339.*)

**India Soybeans: Area Declines from Dry Weather during Planting Season**

USDA forecasts 2014/15 India soybean production at 11 million tons, down 5 percent from last month and unchanged from last year. Area is forecast at a record 11 million hectares, down almost 5 percent from last month, and down 9.8 percent from last year's record level. Yield is forecast at 1.0 tons per hectare, up 11 percent from last year. In the past five years, India's soybean area and yields have averaged 10.43 million hectares and 1.02 tons per hectare, respectively.

India's soybeans are grown exclusively during the *kharif* (southwest monsoon) season under rainfed conditions. Most of the soybean crop is grown in northwest and central India where the main producing states are Madhya Pradesh (53 percent), Maharashtra (34 percent), and Rajasthan (8 percent).

The season started as one of the driest resulting in significant planting delays. Optimum planting is the third week of June to the second week of July, but the window of opportunity for planting extends until early August. (*For more information, please contact Dath Mita at 202-720-7339.*)

## World Agricultural Production
## U.S. Department of Agriculture
Foreign Agricultural Service / Office of Global Analysis
International Production Assessment Division (IPAD / PECAD)
Ag Box 1051, Room 4630, South Building
Washington, DC 20250-1051
http://www.pecad.fas.usda.gov/
Telephone: (202) 720-1157          Fax: (202) 720-1158

This report uses information from the Foreign Agricultural Service's (FAS) global network of agricultural attachés and counselors, official statistics of foreign governments and other foreign source materials, and the analysis of economic data and satellite imagery. Estimates of foreign area, yield, and production are from the International Production Assessment Division, FAS, and are reviewed by USDA's Inter-Agency Commodity Estimates Committee. Estimates of U.S. area, yield, and production are from USDA's National Agricultural Statistics Service. Numbers within the report may not add to totals because of rounding. This report reflects official USDA estimates released in the World Agricultural Supply and Demand Estimates (WASDE-532), August 12, 2014.

**Printed copies are available from the National Technical Information Service. Download an order form at http://www.ntis.gov/products/specialty/usda/fas_a-g.asp, or call NTIS at 1-800-363-2068.**

The FAS International Production Assessment Division prepared this report. The next issue of World Agricultural Production will be released after 12:00 p.m. Eastern Time, September 11, 2014.

### Conversion Table

#### Metric tons to bushels

| | | |
|---|---|---|
| Wheat, soybeans | = | MT * 36.7437 |
| Corn, sorghum, rye | = | MT * 39.36825 |
| Barley | = | MT * 45.929625 |
| Oats | = | MT * 68.894438 |

#### Metric tons to 480-lb bales

| | | |
|---|---|---|
| Cotton | = | MT * 4.592917 |

#### Metric tons to hundredweight

| | | |
|---|---|---|
| Rice | = | MT * 22.04622 |

#### Area & weight

| | | |
|---|---|---|
| 1 hectare | = | 2.471044 acres |
| 1 kilogram | = | 2.204622 pounds |

For further information, contact:
**U.S. Department of Agriculture**
**Foreign Agricultural Service**
**Office of Global Analysis**
International Production Assessment Division
Ag Box 1051, Room 4630, South Building
Washington, DC 20250-1051

Telephone: (202) 720-1157          Fax: (202) 720-1158

---

## GENERAL INFORMATION

| | | | |
|---|---|---|---|
| Acting Director | Paul Provance | 202-720-2974 | paul.provance@fas.usda.gov |
| USDA Remote Sensing Advisor | Glenn Bethel | 202-720-1280 | glenn.bethel@fas.usda.gov |
| Sr. Analyst/Technical Lead | Curt Reynolds, PhD | 202-690-0134 | curt reynolds@fas.usda.gov |
| Sr. Analyst/Satellite Imagery Archive Manager/Technical Lead | Dath Mita, PhD | 202-720-7339 | mita.dath@fas.usda.gov |
| Sr. Analyst/ Global Special Projects Manager/Technical Lead | Jim Crutchfield | 202-690-0135 | james.crutchfield@fas.usda.gov |
| Sr. Analyst/Technical Lead | Robert Tetrault | 202-720-1071 | robert.tetrault@fas.usda.gov |
| GIS Analyst/WAP Coordinator | Justin Jenkins | 202-720-0419 | justin.jenkins@fas.usda.gov |
| Program Analyst | Mary Jackson | 202-720-0886 | mary.jackson@fas.usda.gov |
| Management Analyst | Rishan Chaudhry | 202-720-1157 | rishan.chaudhry@fas.usda.gov |
| Administrative Assistant/COR | Terri Lagarde | 202-720-1156 | terri.lagarde@fas.usda.gov |

## COUNTRY- AND REGION-SPECIFIC INFORMATION

| | | | |
|---|---|---|---|
| South América, Argentina and Colombia | Denise McWilliams, PhD | 202-720-0107 | denise.mcwilliams@fas.usda.gov |
| Western and Central Europe, and North Africa | Bryan Purcell | 202-690-0138 | bryan.purcell@fas.usda.gov |
| Russia, Kazakhstan, Ukraine, and other FSU-12 countries | Mark Lindeman | 202-690-0143 | mark.lindeman@fas.usda.gov |
| Canada, Caribbean, Sri Lanka, and Bangladesh | Arnella Trent | 202-720-0881 | arnella.trent@fas.usda.gov |
| East Asia, China, and Japan | Paulette Sandene | 202-690-0133 | paulette.sandene@fas.usda.gov |
| India, Pakistan, and Nepal | Dath Mita, PhD | 202-720-7339 | mita.dath@fas.usda.gov |
| Sub-Saharan Africa, Mexico, Nigeria and South Africa | Curt Reynolds, PhD | 202-690-0134 | curt reynoldsc@fas.usda.gov |
| S.E. Asia, Indonesia, Thailand, Malaysia, Cambodia, and Vietnam | Michael Shean | 202-720-7366 | michael.shean@fas.usda.gov |
| Brazil, Venezuela, Central America, | Robert Tetrault | 202-720-1071 | robert.tetrault@fas.usda.gov |
| Australia, New Zealand, Papua New Guinea, and South Pacific Islands | Jim Crutchfield | 202-690-0135 | james.crutchfield@fas.usda.gov |
| Middle East, Afghanistan, Iraq, Iran, Syria | Bill Baker, PhD | 202-260-8109 | bill.baker@fas.usda.gov |
| Western United States | Justin Jenkins | 202-720-0419 | justin.jenkins@fas.usda.gov |
| Eastern United States | Paul Provance | 202-720-2974 | paul.provance@fas.usda.gov |
| Crop Analyst | Vacant | | |

The Foreign Agricultural Service (FAS) updates its production, supply and distribution (PSD) database for cotton, oilseeds, and grains at 12:00 p.m. on the day the *World Agricultural Supply and Demand Estimates* (WASDE) report is released. This circular is released by 12:15 p.m.

### FAS Reports and Databases:

Current *World Market* and *Trade and World Agricultural Production* Reports:
http://apps.fas.usda.gov/psdonline/psdDataPublications.aspx
Archives *World Market* and *Trade and World Agricultural Production* Reports:
http://usda.mannlib.cornell.edu/MannUsda/viewTaxonomy.do?taxonomyID=7
Production, Supply and Distribution Database (PSD Online):
http://apps.fas.usda.gov/psdonline/psdHome.aspx
Global Agricultural Trade System (U.S. Exports and Imports):
http://apps.fas.usda.gov/gats/default.aspx
Export Sales Report:
http://apps.fas.usda.gov/esrquery/
Global Agricultural Information Network (Agricultural Attaché Reports):
http://gain.fas.usda.gov/Pages/Default.aspx

### Other USDA Reports:

World Agricultural Supply and Demand Estimates (WASDE):
http://www.usda.gov/oce/commodity/wasde/
Economic Research Service:
http://www.ers.usda.gov/topics/crops
National Agricultural Statistics Service:
http://www.nass.usda.gov/Publications/

## Table 01 World Crop Production Summary

### Million Metric Tons

| Commodity | World | Total Foreign | North America | | | European Union | Former Soviet | | Asia (WAP) | | | | | South America | | Aus-tralia | Selected Other | | All Others |
|---|---|---|---|---|---|---|---|---|---|---|---|---|---|---|---|---|---|---|---|
| | | | United States | Canada | Mexico | | Russia | Ukraine | China | India | Indo-nesia | Paki-stan | Thai-land | Argen-tina | Brazil | | South Africa | Turkey | |
| | | | | | | | | | ---Million metric tons--- | | | | | | | | | | |
| **Wheat** | | | | | | | | | | | | | | | | | | | |
| 2012/13 | 658.2 | 596.5 | 61.7 | 27.2 | 3.2 | 133.9 | 37.7 | 15.8 | 121.0 | 94.9 | nr | 23.3 | nr | 9.3 | 4.4 | 22.9 | 1.9 | 15.5 | 85.6 |
| 2013/14 prel. | 714.1 | 656.1 | 58.0 | 37.5 | 3.4 | 143.1 | 52.1 | 22.3 | 121.9 | 93.5 | nr | 24.0 | nr | 10.5 | 5.3 | 27.0 | 1.8 | 18.0 | 95.7 |
| 2014/15 proj. | | | | | | | | | | | | | | | | | | | |
| Jul | 705.2 | 651.0 | 54.2 | 28.0 | 3.9 | 147.9 | 53.0 | 21.0 | 124.0 | 95.9 | nr | 24.5 | nr | 12.5 | 6.3 | 26.0 | 1.8 | 15.0 | 91.3 |
| Aug | 716.1 | 660.9 | 55.2 | 28.0 | 3.8 | 147.9 | 59.0 | 22.0 | 126.0 | 95.9 | nr | 24.5 | nr | 12.5 | 6.3 | 26.0 | 1.8 | 15.0 | 92.2 |
| **Coarse Grains** | | | | | | | | | | | | | | | | | | | |
| 2012/13 | 1,137.3 | 851.3 | 286.0 | 24.4 | 28.9 | 145.9 | 28.7 | 29.5 | 212.2 | 39.9 | 8.5 | 5.6 | 4.7 | 37.2 | 84.3 | 11.4 | 12.9 | 10.6 | 166.6 |
| 2013/14 prel. | 1,274.3 | 904.9 | 369.4 | 28.7 | 30.2 | 158.2 | 35.7 | 39.9 | 225.1 | 42.7 | 9.1 | 5.6 | 5.0 | 33.5 | 81.1 | 12.4 | 15.3 | 13.1 | 169.3 |
| 2014/15 proj. | | | | | | | | | | | | | | | | | | | |
| Jul | 1,261.2 | 893.3 | 367.9 | 21.8 | 30.1 | 154.0 | 38.5 | 36.1 | 228.6 | 41.7 | 9.2 | 5.6 | 5.0 | 35.2 | 77.1 | 11.8 | 14.0 | 9.6 | 175.0 |
| Aug | 1,267.8 | 894.9 | 372.9 | 21.8 | 30.1 | 155.9 | 40.5 | 36.5 | 228.8 | 39.0 | 9.2 | 5.6 | 5.0 | 35.2 | 77.1 | 11.8 | 14.1 | 9.3 | 175.2 |
| **Rice, Milled** | | | | | | | | | | | | | | | | | | | |
| 2012/13 | 471.7 | 465.4 | 6.3 | nr | 0.1 | 2.1 | 0.7 | 0.1 | 143.0 | 105.2 | 36.6 | 5.8 | 20.2 | 1.0 | 8.0 | 0.8 | nr | 0.5 | 141.2 |
| 2013/14 prel. | 475.8 | 469.7 | 6.1 | nr | 0.1 | 1.9 | 0.6 | 0.1 | 142.5 | 106.3 | 36.0 | 6.6 | 20.5 | 1.0 | 8.3 | 0.6 | nr | 0.5 | 144.6 |
| 2014/15 proj. | | | | | | | | | | | | | | | | | | | |
| Jul | 479.4 | 472.2 | 7.2 | nr | 0.1 | 2.0 | 0.7 | 0.1 | 144.0 | 104.0 | 37.7 | 6.7 | 20.5 | 1.0 | 8.5 | 0.6 | nr | 0.5 | 145.9 |
| Aug | 477.3 | 470.0 | 7.3 | nr | 0.1 | 2.0 | 0.7 | 0.1 | 144.0 | 103.0 | 37.0 | 6.7 | 20.5 | 1.0 | 8.4 | 0.5 | nr | 0.5 | 145.6 |
| **Total Grains** | | | | | | | | | | | | | | | | | | | |
| 2012/13 | 2,267.1 | 1,913.1 | 354.0 | 51.6 | 32.2 | 281.8 | 67.1 | 45.4 | 476.2 | 240.1 | 45.1 | 34.7 | 24.9 | 47.6 | 96.7 | 35.1 | 14.7 | 26.6 | 393.4 |
| 2013/14 prel. | 2,464.2 | 2,030.7 | 433.5 | 66.2 | 33.7 | 303.2 | 88.4 | 62.3 | 489.5 | 242.5 | 45.1 | 36.2 | 25.4 | 45.0 | 94.7 | 40.0 | 17.1 | 31.6 | 409.6 |
| 2014/15 proj. | | | | | | | | | | | | | | | | | | | |
| Jul | 2,445.8 | 2,016.5 | 429.3 | 49.8 | 34.1 | 303.9 | 92.2 | 57.2 | 496.6 | 241.6 | 46.9 | 36.8 | 25.5 | 48.7 | 91.9 | 38.3 | 15.8 | 25.1 | 412.2 |
| Aug | 2,461.2 | 2,025.8 | 435.4 | 49.8 | 34.1 | 305.7 | 100.2 | 58.6 | 498.8 | 237.9 | 46.2 | 36.8 | 25.5 | 48.7 | 91.8 | 38.3 | 15.9 | 24.8 | 413.1 |
| **Oilseeds** | | | | | | | | | | | | | | | | | | | |
| 2012/13 | 474.5 | 381.4 | 93.1 | 19.0 | 0.9 | 28.1 | 10.9 | 12.7 | 59.8 | 36.8 | 10.8 | 5.1 | 0.6 | 53.7 | 84.8 | 5.7 | 1.4 | 2.2 | 48.8 |
| 2013/14 prel. | 503.9 | 406.8 | 97.1 | 23.3 | 0.9 | 31.7 | 13.6 | 16.7 | 58.6 | 38.3 | 11.5 | 5.1 | 0.6 | 57.5 | 91.1 | 5.2 | 1.9 | 2.4 | 48.5 |
| 2014/15 proj. | | | | | | | | | | | | | | | | | | | |
| Jul | 521.9 | 408.8 | 113.1 | 21.9 | 1.0 | 32.3 | 14.4 | 15.5 | 57.8 | 37.8 | 12.1 | 5.0 | 0.6 | 58.5 | 94.6 | 4.4 | 1.9 | 2.3 | 48.8 |
| Aug | 521.8 | 408.1 | 113.7 | 21.4 | 1.0 | 32.6 | 14.0 | 15.7 | 57.8 | 37.6 | 12.1 | 5.0 | 0.6 | 58.5 | 94.3 | 4.4 | 1.9 | 2.3 | 48.8 |
| **Cotton** | | | | | | | | | | | | | | | | | | | |
| 2012/13 | 123.0 | 105.6 | 17.3 | nr | 1.0 | 1.5 | nr | nr | 35.0 | 28.5 | 0.0 | 9.3 | 0.0 | 0.8 | 6.0 | 4.6 | 0.0 | 2.7 | 16.3 |
| 2013/14 prel. | 118.3 | 105.4 | 12.9 | nr | 0.9 | 1.6 | nr | nr | 32.0 | 30.5 | 0.0 | 9.5 | 0.0 | 1.2 | 7.8 | 4.1 | 0.0 | 2.3 | 15.3 |
| 2014/15 proj. | | | | | | | | | | | | | | | | | | | |
| Jul | 116.4 | 99.9 | 16.5 | nr | 1.1 | 1.7 | nr | nr | 29.5 | 28.0 | 0.0 | 9.5 | 0.0 | 1.2 | 8.0 | 2.7 | 0.1 | 2.9 | 15.2 |
| Aug | 117.6 | 100.1 | 17.5 | nr | 1.2 | 1.7 | nr | nr | 29.5 | 29.0 | 0.0 | 9.5 | 0.0 | 1.2 | 7.3 | 2.5 | 0.1 | 2.9 | 15.2 |

1/ Includes wheat, coarse grains, and rice (milled) shown above.

# Table 02 Wheat Area, Yield, and Production

| Country / Region | Area (Million hectares) | | | | Yield (Metric tons per hectare) | | | | Production (Million metric tons) | | | | Change in Production | | | |
|---|---|---|---|---|---|---|---|---|---|---|---|---|---|---|---|---|
| | | | 2014/15 Proj. | | | | 2014/15 Proj. | | | | 2014/15 Proj. | | From last month | | From last year | |
| | 2012/13 | Prel. 2013/14 | Jul | Aug | 2012/13 | Prel. 2013/14 | Jul | Aug | 2012/13 | Prel. 2013/14 | Jul | Aug | MMT | Percent | MMT | Percent |
| **World** | 216.35 | 220.53 | 222.90 | 222.57 | 3.04 | 3.24 | 3.16 | 3.22 | 658.16 | 714.07 | 705.17 | 716.09 | 10.92 | 1.55 | 2.02 | 0.28 |
| **United States** | 19.80 | 18.27 | 18.71 | 18.71 | 3.12 | 3.17 | 2.90 | 2.95 | 61.67 | 57.96 | 54.21 | 55.24 | 1.03 | 1.91 | -2.72 | -4.70 |
| **Total Foreign** | 196.55 | 202.25 | 204.19 | 203.85 | 3.03 | 3.24 | 3.19 | 3.24 | 596.49 | 656.11 | 650.97 | 660.85 | 9.88 | 1.52 | 4.74 | 0.72 |
| **China** | 24.27 | 24.12 | 24.20 | 24.10 | 4.99 | 5.06 | 5.12 | 5.23 | 121.02 | 121.93 | 124.00 | 126.00 | 2.00 | 1.61 | 4.07 | 3.34 |
| **South Asia** | | | | | | | | | | | | | | | | |
| India | 29.86 | 30.00 | 30.60 | 30.60 | 3.18 | 3.12 | 3.13 | 3.13 | 94.88 | 93.51 | 95.85 | 95.85 | 0.00 | 0.00 | 2.34 | 2.50 |
| Pakistan | 8.66 | 8.64 | 8.83 | 8.83 | 2.69 | 2.78 | 2.77 | 2.77 | 23.30 | 24.00 | 24.50 | 24.50 | 0.00 | 0.00 | 0.50 | 2.08 |
| Afghanistan | 2.51 | 2.55 | 2.56 | 2.56 | 2.01 | 1.96 | 1.96 | 1.96 | 5.05 | 5.00 | 5.03 | 5.03 | 0.00 | 0.00 | 0.03 | 0.50 |
| **Former Soviet Union - 12** | | | | | | | | | | | | | | | | |
| Russia | 21.30 | 23.40 | 23.75 | 23.75 | 1.77 | 2.23 | 2.23 | 2.48 | 37.72 | 52.09 | 53.00 | 59.00 | 6.00 | 11.32 | 6.91 | 13.26 |
| Ukraine | 5.63 | 6.57 | 6.30 | 6.30 | 2.80 | 3.39 | 3.33 | 3.49 | 15.76 | 22.28 | 21.00 | 22.00 | 1.00 | 4.76 | -0.28 | -1.25 |
| Kazakhstan | 12.40 | 12.95 | 12.70 | 12.70 | 0.79 | 1.08 | 1.06 | 1.06 | 9.84 | 13.94 | 13.50 | 13.50 | 0.00 | 0.00 | -0.44 | -3.16 |
| Uzbekistan | 1.40 | 1.40 | 1.40 | 1.40 | 4.79 | 4.86 | 4.86 | 4.86 | 6.70 | 6.80 | 6.80 | 6.80 | 0.00 | 0.00 | 0.00 | 0.00 |
| **European Union - 28** | 25.95 | 25.77 | 26.57 | 26.57 | 5.16 | 5.55 | 5.57 | 5.57 | 133.85 | 143.13 | 147.88 | 147.87 | -0.01 | 0.00 | 4.73 | 3.31 |
| France | 5.30 | 5.32 | 5.31 | 5.31 | 7.15 | 7.24 | 7.26 | 7.30 | 37.89 | 38.51 | 38.50 | 38.70 | 0.20 | 0.52 | 0.19 | 0.50 |
| Germany | 3.06 | 3.13 | 3.22 | 3.22 | 7.33 | 7.98 | 7.95 | 7.95 | 22.41 | 24.93 | 25.60 | 25.60 | 0.00 | 0.00 | 0.67 | 2.70 |
| United Kingdom | 1.99 | 1.63 | 1.98 | 1.98 | 6.66 | 7.33 | 7.75 | 7.80 | 13.26 | 11.92 | 15.30 | 15.40 | 0.10 | 0.65 | 3.48 | 29.18 |
| Poland | 2.08 | 2.14 | 2.20 | 2.20 | 4.14 | 4.45 | 4.36 | 4.36 | 8.61 | 9.50 | 9.60 | 9.60 | 0.00 | 0.00 | 0.10 | 1.05 |
| Spain | 2.17 | 2.12 | 2.15 | 2.15 | 2.35 | 3.58 | 2.89 | 2.89 | 5.09 | 7.60 | 6.20 | 6.20 | 0.00 | 0.00 | -1.40 | -18.40 |
| Italy | 1.87 | 1.86 | 1.85 | 1.84 | 4.09 | 3.88 | 4.01 | 3.86 | 7.63 | 7.22 | 7.40 | 7.10 | -0.30 | -4.05 | -0.12 | -1.66 |
| Denmark | 0.62 | 0.57 | 0.70 | 0.70 | 7.42 | 7.29 | 7.27 | 7.48 | 4.56 | 4.14 | 5.05 | 5.20 | 0.15 | 2.97 | 1.06 | 25.57 |
| Hungary | 1.06 | 1.09 | 1.10 | 1.10 | 3.74 | 4.62 | 4.36 | 4.41 | 3.97 | 5.04 | 4.80 | 4.85 | 0.05 | 1.04 | -0.19 | -3.77 |
| Romania | 1.99 | 2.10 | 2.08 | 2.08 | 2.66 | 3.57 | 3.57 | 3.57 | 5.30 | 7.50 | 7.40 | 7.40 | 0.00 | 0.00 | -0.10 | -1.33 |
| Bulgaria | 1.19 | 1.19 | 1.17 | 1.17 | 3.76 | 4.12 | 4.10 | 4.18 | 4.46 | 4.90 | 4.80 | 4.90 | 0.10 | 2.08 | 0.00 | 0.00 |
| **Canada** | 9.50 | 10.44 | 9.60 | 9.30 | 2.86 | 3.59 | 2.92 | 3.01 | 27.21 | 37.50 | 28.00 | 28.00 | 0.00 | 0.00 | -9.50 | -25.33 |
| **Australia** | 12.98 | 13.51 | 13.80 | 13.80 | 1.76 | 2.00 | 1.88 | 1.88 | 22.86 | 27.01 | 26.00 | 26.00 | 0.00 | 0.00 | -1.01 | -3.75 |
| **Middle East** | | | | | | | | | | | | | | | | |
| Turkey | 7.80 | 7.70 | 7.71 | 7.71 | 1.99 | 2.34 | 1.95 | 1.95 | 15.50 | 18.00 | 15.00 | 15.00 | 0.00 | 0.00 | -3.00 | -16.67 |
| Iran | 7.00 | 7.00 | 6.80 | 6.80 | 1.97 | 2.07 | 1.91 | 1.91 | 13.80 | 14.50 | 13.00 | 13.00 | 0.00 | 0.00 | -1.50 | -10.34 |
| Syria | 1.60 | 1.55 | 1.30 | 1.30 | 2.31 | 2.58 | 1.92 | 1.92 | 3.70 | 4.00 | 2.50 | 2.50 | 0.00 | 0.00 | -1.50 | -37.50 |
| **North Africa** | | | | | | | | | | | | | | | | |
| Egypt | 1.35 | 1.35 | 1.40 | 1.40 | 6.30 | 6.41 | 6.39 | 6.39 | 8.50 | 8.65 | 8.95 | 8.95 | 0.00 | 0.00 | 0.30 | 3.47 |
| Morocco | 3.14 | 3.28 | 3.06 | 3.06 | 1.23 | 2.13 | 1.54 | 1.54 | 3.87 | 7.00 | 4.70 | 4.70 | 0.00 | 0.00 | -2.30 | -32.86 |
| **Argentina** | 3.60 | 3.50 | 4.20 | 4.20 | 2.58 | 3.00 | 2.98 | 2.98 | 9.30 | 10.50 | 12.50 | 12.50 | 0.00 | 0.00 | 2.00 | 19.05 |
| **Others** | 17.61 | 18.53 | 19.41 | 19.48 | 2.48 | 2.50 | 2.51 | 2.55 | 43.63 | 46.26 | 48.77 | 49.66 | 0.89 | 1.82 | 3.39 | 7.34 |

World and Selected Countries and Regions

# Table 03 Total Coarse Grain Area, Yield, and Production

| Country / Region | Area (Million hectares) | | | | Yield (Metric tons per hectare) | | | | Production (Million metric tons) | | | | Change in Production | | | |
|---|---|---|---|---|---|---|---|---|---|---|---|---|---|---|---|---|
| | | | 2014/15 Proj. | | | | 2014/15 Proj. | | | | 2014/15 Proj. | | From last month | | From last year | |
| | 2012/13 | Prel. 2013/14 | Jul | Aug | 2012/13 | Prel. 2013/14 | Jul | Aug | 2012/13 | Prel. 2013/14 | Jul | Aug | MMT | Percent | MMT | Percent |
| **World** | 315.68 | 320.71 | 319.35 | 317.56 | 3.60 | 3.97 | 3.95 | 3.99 | 1,137.26 | 1,274.30 | 1,261.17 | 1,267.81 | 6.64 | 0.53 | -6.49 | -0.51 |
| **United States** | 39.20 | 39.87 | 38.18 | 38.18 | 7.30 | 9.27 | 9.64 | 9.77 | 286.01 | 369.43 | 367.86 | 372.86 | 5.00 | 1.36 | 3.43 | 0.93 |
| **Total Foreign** | 276.47 | 280.84 | 281.18 | 279.38 | 3.08 | 3.22 | 3.18 | 3.20 | 851.25 | 904.87 | 893.30 | 894.95 | 1.64 | 0.18 | -9.92 | -1.10 |
| **China** | 37.08 | 38.38 | 38.85 | 38.87 | 5.72 | 5.86 | 5.88 | 5.89 | 212.19 | 225.07 | 228.58 | 228.75 | 0.17 | 0.07 | 3.68 | 1.64 |
| **South America** | | | | | | | | | | | | | | | | |
| Brazil | 16.88 | 16.63 | 15.98 | 15.98 | 4.99 | 4.88 | 4.83 | 4.83 | 84.25 | 81.11 | 77.11 | 77.11 | 0.00 | 0.00 | -4.00 | -4.93 |
| Argentina | 6.84 | 5.93 | 6.02 | 6.02 | 5.45 | 5.64 | 5.85 | 5.85 | 37.25 | 33.47 | 35.17 | 35.17 | 0.00 | 0.00 | 1.70 | 5.08 |
| **European Union - 28** | 31.34 | 31.40 | 30.79 | 30.86 | 4.66 | 5.04 | 5.00 | 5.05 | 145.89 | 158.16 | 154.05 | 155.89 | 1.84 | 1.20 | -2.27 | -1.44 |
| France | 3.99 | 4.04 | 4.06 | 4.09 | 7.50 | 6.98 | 7.38 | 7.45 | 29.94 | 28.23 | 29.95 | 30.45 | 0.50 | 1.67 | 2.22 | 7.86 |
| Germany | 3.46 | 3.40 | 3.29 | 3.29 | 6.64 | 6.60 | 6.94 | 6.94 | 22.99 | 22.45 | 22.84 | 22.84 | 0.00 | 0.00 | 0.39 | 1.73 |
| Poland | 5.53 | 5.21 | 5.16 | 5.17 | 3.58 | 3.59 | 3.53 | 3.54 | 19.80 | 18.70 | 18.19 | 18.29 | 0.11 | 0.58 | -0.41 | -2.19 |
| Spain | 3.79 | 3.95 | 3.89 | 3.92 | 3.01 | 4.23 | 3.23 | 3.17 | 11.41 | 16.69 | 12.56 | 12.41 | -0.16 | -1.27 | -4.29 | -25.68 |
| Italy | 1.40 | 1.24 | 1.21 | 1.22 | 6.47 | 6.70 | 7.32 | 7.38 | 9.08 | 8.31 | 8.88 | 9.03 | 0.15 | 1.69 | 0.73 | 8.73 |
| Hungary | 1.67 | 1.72 | 1.77 | 1.77 | 3.78 | 4.95 | 5.19 | 5.30 | 6.30 | 8.53 | 9.21 | 9.41 | 0.20 | 2.17 | 0.88 | 10.28 |
| United Kingdom | 1.15 | 1.42 | 1.15 | 1.16 | 5.44 | 5.77 | 5.83 | 5.83 | 6.25 | 8.17 | 6.68 | 6.73 | 0.06 | 0.82 | -1.44 | -17.64 |
| Romania | 3.40 | 3.42 | 3.35 | 3.35 | 2.23 | 3.72 | 3.47 | 3.63 | 7.59 | 12.75 | 11.61 | 12.16 | 0.55 | 4.74 | -0.59 | -4.61 |
| **Former Soviet Union - 12** | | | | | | | | | | | | | | | | |
| Russia | 14.18 | 15.49 | 16.20 | 16.20 | 2.02 | 2.31 | 2.38 | 2.50 | 28.66 | 35.74 | 38.50 | 40.50 | 2.00 | 5.19 | 4.77 | 13.33 |
| Ukraine | 8.55 | 8.78 | 8.47 | 8.32 | 3.45 | 4.55 | 4.26 | 4.39 | 29.53 | 39.92 | 36.10 | 36.51 | 0.41 | 1.12 | -3.41 | -8.55 |
| Kazakhstan | 1.95 | 2.26 | 2.39 | 2.39 | 1.17 | 1.55 | 1.44 | 1.44 | 2.28 | 3.51 | 3.44 | 3.44 | 0.00 | 0.00 | -0.07 | -2.02 |
| **Africa** | 78.07 | 79.20 | 81.08 | 81.08 | 1.44 | 1.39 | 1.41 | 1.41 | 112.55 | 110.46 | 114.38 | 114.40 | 0.02 | 0.01 | 3.94 | 3.57 |
| Nigeria | 12.73 | 13.25 | 13.25 | 13.25 | 1.46 | 1.45 | 1.45 | 1.45 | 18.57 | 19.20 | 19.20 | 19.20 | 0.00 | 0.00 | 0.00 | 0.00 |
| South Africa | 3.44 | 3.31 | 3.41 | 3.41 | 3.75 | 4.63 | 4.12 | 4.12 | 12.87 | 15.33 | 14.04 | 14.06 | 0.02 | 0.11 | -1.27 | -8.30 |
| Ethiopia | 5.18 | 5.41 | 5.41 | 5.41 | 2.37 | 2.41 | 2.41 | 2.41 | 12.29 | 13.05 | 13.05 | 13.05 | 0.00 | 0.00 | 0.00 | 0.00 |
| Egypt | 0.98 | 0.94 | 0.94 | 0.94 | 6.83 | 7.09 | 7.07 | 7.07 | 6.67 | 6.67 | 6.62 | 6.62 | 0.00 | 0.00 | -0.05 | -0.75 |
| **India** | 25.08 | 25.38 | 25.51 | 23.91 | 1.59 | 1.68 | 1.64 | 1.63 | 39.94 | 42.71 | 41.73 | 39.03 | -2.70 | -6.47 | -3.68 | -8.62 |
| **Southeast Asia** | | | | | | | | | | | | | | | | |
| Indonesia | 3.00 | 3.12 | 3.12 | 3.12 | 2.83 | 2.92 | 2.95 | 2.95 | 8.50 | 9.10 | 9.20 | 9.20 | 0.00 | 0.00 | 0.10 | 1.10 |
| Philippines | 2.56 | 2.59 | 2.63 | 2.63 | 2.84 | 2.92 | 3.01 | 3.01 | 7.26 | 7.54 | 7.90 | 7.90 | 0.00 | 0.00 | 0.36 | 4.77 |
| Thailand | 1.11 | 1.15 | 1.13 | 1.13 | 4.20 | 4.31 | 4.38 | 4.38 | 4.66 | 4.96 | 4.96 | 4.96 | 0.00 | 0.00 | 0.00 | 0.00 |
| **Mexico** | 8.92 | 9.15 | 9.00 | 9.00 | 3.24 | 3.30 | 3.35 | 3.35 | 28.88 | 30.21 | 30.14 | 30.14 | 0.00 | 0.00 | -0.07 | -0.23 |
| **Canada** | 5.35 | 5.39 | 4.51 | 4.51 | 4.57 | 5.33 | 4.83 | 4.83 | 24.43 | 28.74 | 21.77 | 21.77 | 0.00 | 0.00 | -6.97 | -24.26 |
| **Australia** | 5.20 | 5.29 | 5.34 | 5.34 | 2.20 | 2.34 | 2.20 | 2.20 | 11.41 | 12.37 | 11.76 | 11.76 | 0.00 | 0.00 | -0.61 | -4.96 |
| **Turkey** | 4.16 | 4.24 | 4.28 | 4.28 | 2.55 | 3.08 | 2.24 | 2.17 | 10.60 | 13.08 | 9.58 | 9.28 | -0.30 | -3.13 | -3.80 | -29.06 |
| **Others** | 26.23 | 26.49 | 25.91 | 25.78 | 2.40 | 2.59 | 2.66 | 2.68 | 62.99 | 68.74 | 68.96 | 69.17 | 0.21 | 0.30 | 0.42 | 0.62 |

World and Selected Countries and Regions

# Table 04 Corn Area, Yield, and Production

| Country / Region | Area (Million hectares) | | | | Yield (Metric tons per hectare) | | | | Production (Million metric tons) | | | | Change in Production | | | |
|---|---|---|---|---|---|---|---|---|---|---|---|---|---|---|---|---|
| | | | 2014/15 Proj. | | | | 2014/15 Proj. | | | | 2014/15 Proj. | | From last month | | From last year | |
| | 2012/13 | Prel. 2013/14 | Jul | Aug | 2012/13 | Prel. 2013/14 | Jul | Aug | 2012/13 | Prel. 2013/14 | Jul | Aug | MMT | Percent | MMT | Percent |
| **World** | 177.23 | 178.03 | 177.28 | 176.81 | 4.90 | 5.53 | 5.53 | 5.57 | 868.76 | 984.37 | 980.96 | 985.39 | 4.43 | 0.45 | 1.02 | 0.10 |
| **United States** | 35.36 | 35.48 | 33.93 | 33.93 | 7.74 | 9.97 | 10.38 | 10.51 | 273.83 | 353.72 | 352.06 | 356.43 | 4.37 | 1.24 | 2.71 | 0.77 |
| **Total Foreign** | 141.87 | 142.55 | 143.35 | 142.88 | 4.19 | 4.42 | 4.39 | 4.40 | 594.93 | 630.66 | 628.90 | 628.97 | 0.06 | 0.01 | -1.69 | -0.27 |
| **China** | 35.03 | 36.32 | 36.80 | 36.80 | 5.87 | 6.02 | 6.03 | 6.03 | 205.61 | 218.49 | 222.00 | 222.00 | 0.00 | 0.00 | 3.51 | 1.61 |
| **South America** | 22.71 | 21.72 | 21.48 | 21.49 | 5.27 | 5.19 | 5.20 | 5.20 | 119.77 | 112.71 | 111.67 | 111.63 | -0.04 | -0.04 | -1.08 | -0.96 |
| Brazil | 15.80 | 15.50 | 14.80 | 14.80 | 5.16 | 5.03 | 5.00 | 5.00 | 81.50 | 78.00 | 74.00 | 74.00 | 0.00 | 0.00 | -4.00 | -5.13 |
| Argentina | 4.00 | 3.40 | 3.65 | 3.65 | 6.75 | 7.06 | 7.12 | 7.12 | 27.00 | 24.00 | 26.00 | 26.00 | 0.00 | 0.00 | 2.00 | 8.33 |
| Bolivia | 0.31 | 0.32 | 0.32 | 0.32 | 2.26 | 2.30 | 2.30 | 2.30 | 0.70 | 0.73 | 0.73 | 0.73 | 0.00 | 0.00 | 0.00 | 0.00 |
| **Mexico** | 6.90 | 6.90 | 6.90 | 6.90 | 3.13 | 3.25 | 3.26 | 3.26 | 21.59 | 22.40 | 22.50 | 22.50 | 0.00 | 0.00 | 0.10 | 0.45 |
| **European Union - 28** | 9.72 | 9.79 | 9.54 | 9.56 | 6.06 | 6.54 | 6.88 | 7.02 | 58.87 | 63.99 | 65.64 | 67.05 | 1.40 | 2.14 | 3.06 | 4.78 |
| France | 1.64 | 1.75 | 1.70 | 1.70 | 9.22 | 8.38 | 9.29 | 9.35 | 15.15 | 14.70 | 15.80 | 15.90 | 0.10 | 0.63 | 1.20 | 8.16 |
| Italy | 0.97 | 0.90 | 0.85 | 0.86 | 7.82 | 7.90 | 8.88 | 8.95 | 7.59 | 7.11 | 7.55 | 7.70 | 0.15 | 1.99 | 0.59 | 8.28 |
| Hungary | 1.19 | 1.25 | 1.26 | 1.26 | 3.98 | 5.36 | 5.79 | 5.95 | 4.74 | 6.73 | 7.30 | 7.50 | 0.20 | 2.74 | 0.78 | 11.52 |
| Romania | 2.73 | 2.65 | 2.55 | 2.55 | 2.23 | 4.00 | 3.63 | 3.84 | 6.10 | 10.60 | 9.25 | 9.80 | 0.55 | 5.95 | -0.80 | -7.55 |
| Poland | 0.54 | 0.61 | 0.56 | 0.58 | 7.35 | 6.51 | 6.73 | 6.74 | 4.00 | 4.00 | 3.77 | 3.88 | 0.11 | 2.79 | -0.13 | -3.13 |
| **India** | 8.91 | 9.50 | 9.00 | 8.60 | 2.50 | 2.55 | 2.44 | 2.44 | 22.26 | 24.19 | 22.00 | 21.00 | -1.00 | -4.55 | -3.19 | -13.19 |
| **Canada** | 1.42 | 1.48 | 1.23 | 1.23 | 9.21 | 9.59 | 9.43 | 9.43 | 13.06 | 14.20 | 11.60 | 11.60 | 0.00 | 0.00 | -2.60 | -18.31 |
| **Indonesia** | 3.00 | 3.12 | 3.12 | 3.12 | 2.83 | 2.92 | 2.95 | 2.95 | 8.50 | 9.10 | 9.20 | 9.20 | 0.00 | 0.00 | 0.10 | 1.10 |
| **Ukraine** | 4.37 | 4.83 | 4.80 | 4.70 | 4.79 | 6.40 | 5.63 | 5.74 | 20.92 | 30.90 | 27.00 | 27.00 | 0.00 | 0.00 | -3.90 | -12.62 |
| **Serbia** | 1.30 | 1.25 | 1.28 | 1.28 | 2.88 | 5.12 | 5.18 | 5.18 | 3.75 | 6.40 | 6.60 | 6.60 | 0.00 | 0.00 | 0.20 | 3.13 |
| **Egypt** | 0.75 | 0.71 | 0.71 | 0.71 | 7.73 | 8.12 | 8.10 | 8.10 | 5.80 | 5.80 | 5.75 | 5.75 | 0.00 | 0.00 | -0.05 | -0.86 |
| **Philippines** | 2.56 | 2.59 | 2.63 | 2.63 | 2.84 | 2.92 | 3.01 | 3.01 | 7.26 | 7.54 | 7.90 | 7.90 | 0.00 | 0.00 | 0.36 | 4.77 |
| **Vietnam** | 1.16 | 1.17 | 1.20 | 1.20 | 4.15 | 4.43 | 4.50 | 4.50 | 4.80 | 5.20 | 5.40 | 5.40 | 0.00 | 0.00 | 0.20 | 3.93 |
| **Thailand** | 1.08 | 1.12 | 1.10 | 1.10 | 4.26 | 4.38 | 4.45 | 4.45 | 4.60 | 4.90 | 4.90 | 4.90 | 0.00 | 0.00 | 0.00 | 0.00 |
| **Russia** | 1.94 | 2.32 | 2.60 | 2.60 | 4.24 | 5.01 | 5.00 | 5.00 | 8.21 | 11.64 | 13.00 | 13.00 | 0.00 | 0.00 | 1.37 | 11.73 |
| **Sub-Saharan Africa** | | | | | | | | | | | | | | | | |
| South Africa | 3.24 | 3.10 | 3.20 | 3.20 | 3.82 | 4.76 | 4.22 | 4.22 | 12.37 | 14.75 | 13.50 | 13.50 | 0.00 | 0.00 | -1.25 | -8.47 |
| Nigeria | 4.16 | 4.25 | 4.25 | 4.25 | 1.83 | 1.81 | 1.81 | 1.81 | 7.63 | 7.70 | 7.70 | 7.70 | 0.00 | 0.00 | 0.00 | 0.00 |
| Ethiopia | 2.01 | 2.15 | 2.15 | 2.15 | 3.06 | 3.02 | 3.02 | 3.02 | 6.16 | 6.50 | 6.50 | 6.50 | 0.00 | 0.00 | 0.00 | 0.00 |
| Zimbabwe | 0.96 | 0.95 | 1.30 | 1.30 | 1.01 | 0.84 | 1.00 | 1.00 | 0.97 | 0.80 | 1.30 | 1.30 | 0.00 | 0.00 | 0.50 | 62.50 |
| **Turkey** | 0.53 | 0.58 | 0.55 | 0.55 | 8.38 | 8.79 | 8.91 | 8.36 | 4.40 | 5.10 | 4.90 | 4.60 | -0.30 | -6.12 | -0.50 | -9.80 |
| **Others** | 30.14 | 28.70 | 29.53 | 29.53 | 1.94 | 2.03 | 2.03 | 2.03 | 58.40 | 58.35 | 59.84 | 59.84 | 0.00 | 0.00 | 1.49 | 2.55 |

World and Selected Countries and Regions

# Table 05 Barley Area, Yield, and Production

| Country / Region | Area (Million hectares) | | | | Yield (Metric tons per hectare) | | | | Production (Million metric tons) | | | | Change in Production | | | |
|---|---|---|---|---|---|---|---|---|---|---|---|---|---|---|---|---|
| | 2012/13 | Prel. 2013/14 | 2014/15 Proj. Jul | 2014/15 Proj. Aug | 2012/13 | Prel. 2013/14 | 2014/15 Proj. Jul | 2014/15 Proj. Aug | 2012/13 | Prel. 2013/14 | 2014/15 Proj. Jul | 2014/15 Proj. Aug | From last month MMT | Percent | From last year MMT | Percent |
| **World** | 50.18 | 50.52 | 48.96 | 49.19 | 2.59 | 2.87 | 2.69 | 2.76 | 129.84 | 145.05 | 131.94 | 135.65 | 3.71 | 2.82 | -9.40 | -6.48 |
| **United States** | 1.31 | 1.21 | 1.07 | 1.07 | 3.65 | 3.86 | 3.83 | 3.94 | 4.80 | 4.68 | 4.08 | 4.20 | 0.12 | 2.84 | -0.49 | -10.40 |
| **Total Foreign** | 48.87 | 49.30 | 47.90 | 48.13 | 2.56 | 2.85 | 2.67 | 2.73 | 125.04 | 140.37 | 127.86 | 131.45 | 3.60 | 2.81 | -8.92 | -6.35 |
| **Russia** | 7.63 | 8.02 | 8.20 | 8.20 | 1.83 | 1.92 | 2.01 | 2.26 | 13.95 | 15.39 | 16.50 | 18.50 | 2.00 | 12.12 | 3.11 | 20.22 |
| **European Union - 28** | 12.51 | 12.35 | 12.26 | 12.32 | 4.39 | 4.83 | 4.56 | 4.56 | 54.90 | 59.61 | 55.87 | 56.20 | 0.33 | 0.60 | -3.41 | -5.72 |
| Germany | 1.68 | 1.57 | 1.60 | 1.60 | 6.19 | 6.59 | 6.61 | 6.61 | 10.39 | 10.37 | 10.58 | 10.58 | 0.00 | 0.00 | 0.21 | 2.05 |
| France | 1.68 | 1.63 | 1.70 | 1.73 | 6.74 | 6.33 | 6.41 | 6.53 | 11.34 | 10.34 | 10.90 | 11.30 | 0.40 | 3.67 | 0.96 | 9.28 |
| Spain | 2.68 | 2.77 | 2.76 | 2.76 | 2.23 | 3.63 | 2.54 | 2.43 | 5.98 | 10.06 | 7.00 | 6.70 | -0.30 | -4.29 | -3.36 | -33.40 |
| United Kingdom | 1.00 | 1.21 | 1.00 | 1.00 | 5.51 | 5.84 | 5.90 | 5.90 | 5.52 | 7.09 | 5.90 | 5.90 | 0.00 | 0.00 | -1.19 | -16.81 |
| Denmark | 0.72 | 0.69 | 0.60 | 0.60 | 5.56 | 5.78 | 5.67 | 5.67 | 4.01 | 3.98 | 3.40 | 3.40 | 0.00 | 0.00 | -0.58 | -14.62 |
| Poland | 1.16 | 0.82 | 0.90 | 0.90 | 3.60 | 3.55 | 3.53 | 3.53 | 4.18 | 2.90 | 3.18 | 3.18 | 0.00 | 0.00 | 0.28 | 9.66 |
| Czech Republic | 0.38 | 0.35 | 0.35 | 0.35 | 4.23 | 4.57 | 4.71 | 4.74 | 1.62 | 1.59 | 1.65 | 1.66 | 0.01 | 0.61 | 0.07 | 4.14 |
| Finland | 0.45 | 0.50 | 0.50 | 0.50 | 3.50 | 3.90 | 3.60 | 3.60 | 1.59 | 1.93 | 1.80 | 1.80 | 0.00 | 0.00 | -0.13 | -6.93 |
| Sweden | 0.37 | 0.39 | 0.33 | 0.33 | 4.61 | 5.01 | 4.70 | 4.70 | 1.70 | 1.93 | 1.55 | 1.55 | 0.00 | 0.00 | -0.38 | -19.81 |
| Italy | 0.25 | 0.19 | 0.22 | 0.22 | 3.82 | 3.62 | 3.86 | 3.86 | 0.94 | 0.68 | 0.85 | 0.85 | 0.00 | 0.00 | 0.17 | 24.27 |
| Hungary | 0.28 | 0.27 | 0.30 | 0.30 | 3.61 | 4.06 | 4.00 | 4.00 | 1.00 | 1.08 | 1.20 | 1.20 | 0.00 | 0.00 | 0.12 | 11.11 |
| Austria | 0.15 | 0.14 | 0.15 | 0.15 | 4.38 | 5.13 | 5.17 | 5.17 | 0.66 | 0.73 | 0.75 | 0.75 | 0.00 | 0.00 | 0.02 | 2.18 |
| **Ukraine** | 3.29 | 3.23 | 3.00 | 3.00 | 2.11 | 2.34 | 2.60 | 2.73 | 6.94 | 7.56 | 7.80 | 8.20 | 0.40 | 5.13 | 0.64 | 8.45 |
| **Canada** | 2.75 | 2.65 | 2.20 | 2.20 | 2.91 | 3.87 | 3.18 | 3.18 | 8.01 | 10.25 | 7.00 | 7.00 | 0.00 | 0.00 | -3.25 | -31.71 |
| **Australia** | 3.64 | 3.92 | 3.80 | 3.80 | 2.05 | 2.43 | 2.13 | 2.13 | 7.47 | 9.55 | 8.10 | 8.10 | 0.00 | 0.00 | -1.45 | -15.14 |
| **Turkey** | 3.30 | 3.33 | 3.40 | 3.40 | 1.67 | 2.19 | 1.18 | 1.18 | 5.50 | 7.30 | 4.00 | 4.00 | 0.00 | 0.00 | -3.30 | -45.21 |
| **China** | 0.49 | 0.45 | 0.43 | 0.45 | 3.32 | 3.33 | 3.26 | 3.44 | 1.63 | 1.50 | 1.40 | 1.55 | 0.15 | 10.71 | 0.05 | 3.33 |
| **Iran** | 1.68 | 1.58 | 1.58 | 1.58 | 2.03 | 2.03 | 2.03 | 2.03 | 3.40 | 3.20 | 3.20 | 3.20 | 0.00 | 0.00 | 0.00 | 0.00 |
| **Morocco** | 1.89 | 1.69 | 1.44 | 1.44 | 0.63 | 1.60 | 1.22 | 1.22 | 1.20 | 2.70 | 1.75 | 1.75 | 0.00 | 0.00 | -0.95 | -35.19 |
| **Kazakhstan** | 1.60 | 1.84 | 2.00 | 2.00 | 0.94 | 1.38 | 1.30 | 1.30 | 1.50 | 2.54 | 2.60 | 2.60 | 0.00 | 0.00 | 0.06 | 2.40 |
| **Ethiopia** | 1.02 | 1.02 | 1.02 | 1.02 | 1.75 | 1.76 | 1.76 | 1.76 | 1.78 | 1.80 | 1.80 | 1.80 | 0.00 | 0.00 | 0.00 | 0.00 |
| **Belarus** | 0.56 | 0.57 | 0.45 | 0.60 | 3.44 | 2.94 | 3.11 | 3.50 | 1.92 | 1.67 | 1.40 | 2.10 | 0.70 | 50.00 | 0.43 | 25.45 |
| **India** | 0.77 | 0.78 | 0.81 | 0.81 | 2.10 | 2.24 | 2.14 | 2.14 | 1.62 | 1.75 | 1.73 | 1.73 | 0.00 | 0.00 | -0.02 | -1.14 |
| **Argentina** | 1.50 | 1.27 | 1.00 | 1.00 | 3.33 | 3.74 | 3.85 | 3.85 | 5.00 | 4.75 | 3.85 | 3.85 | 0.00 | 0.00 | -0.90 | -18.95 |
| **Mexico** | 0.33 | 0.22 | 0.22 | 0.22 | 3.15 | 2.56 | 2.56 | 2.56 | 1.03 | 0.55 | 0.55 | 0.55 | 0.00 | 0.00 | 0.00 | 0.00 |
| **Iraq** | 0.60 | 1.07 | 1.15 | 1.15 | 0.83 | 0.88 | 0.87 | 0.87 | 0.50 | 0.94 | 1.00 | 1.00 | 0.00 | 0.00 | 0.06 | 5.93 |
| **Algeria** | 1.00 | 0.90 | 1.00 | 1.00 | 1.50 | 1.66 | 1.65 | 1.65 | 1.50 | 1.50 | 1.65 | 1.65 | 0.00 | 0.00 | 0.15 | 10.15 |
| **Others** | 4.31 | 4.42 | 3.94 | 3.95 | 1.67 | 1.77 | 1.94 | 1.94 | 7.20 | 7.81 | 7.66 | 7.67 | 0.02 | 0.20 | -0.14 | -1.74 |

World and Selected Countries and Regions

# Table 06 Oats Area, Yield, and Production

| Country / Region | Area (Million hectares) | | | | Yield (Metric tons per hectare) | | | | Production (Million metric tons) | | | | Change in Production | | | |
|---|---|---|---|---|---|---|---|---|---|---|---|---|---|---|---|---|
| | 2012/13 | Prel. 2013/14 | 2014/15 Proj. Jul | 2014/15 Proj. Aug | 2012/13 | Prel. 2013/14 | 2014/15 Proj. Jul | 2014/15 Proj. Aug | 2012/13 | Prel. 2013/14 | 2014/15 Proj. Jul | 2014/15 Proj. Aug | From last month MMT | From last month Percent | From last year MMT | From last year Percent |
| **World** | 9.46 | 9.66 | 9.56 | 9.55 | 2.23 | 2.44 | 2.33 | 2.36 | 21.14 | 23.58 | 22.27 | 22.52 | 0.25 | 1.11 | -1.06 | -4.51 |
| **United States** | 0.42 | 0.42 | 0.47 | 0.47 | 2.20 | 2.29 | 2.35 | 2.40 | 0.93 | 0.96 | 1.10 | 1.12 | 0.03 | 2.37 | 0.17 | 17.36 |
| **Total Foreign** | 9.04 | 9.24 | 9.10 | 9.08 | 2.24 | 2.45 | 2.33 | 2.36 | 20.21 | 22.62 | 21.17 | 21.39 | 0.22 | 1.05 | -1.23 | -5.43 |
| **Russia** | 2.86 | 3.01 | 3.00 | 3.00 | 1.41 | 1.64 | 1.67 | 1.67 | 4.03 | 4.93 | 5.00 | 5.00 | 0.00 | 0.00 | 0.07 | 1.38 |
| **Canada** | 0.99 | 1.11 | 0.94 | 0.94 | 2.85 | 3.50 | 2.97 | 2.97 | 2.81 | 3.89 | 2.80 | 2.80 | 0.00 | 0.00 | -1.09 | -28.02 |
| **European Union - 28** | 2.67 | 2.62 | 2.62 | 2.62 | 2.96 | 3.20 | 3.05 | 3.09 | 7.91 | 8.38 | 7.98 | 8.08 | 0.10 | 1.28 | -0.30 | -3.56 |
| Poland | 0.51 | 0.43 | 0.45 | 0.45 | 2.86 | 2.76 | 2.70 | 2.70 | 1.47 | 1.20 | 1.22 | 1.22 | 0.00 | 0.00 | 0.02 | 1.25 |
| Finland | 0.31 | 0.35 | 0.36 | 0.36 | 3.40 | 3.50 | 3.38 | 3.38 | 1.07 | 1.23 | 1.20 | 1.20 | 0.00 | 0.00 | -0.03 | -2.36 |
| Spain | 0.44 | 0.43 | 0.44 | 0.44 | 1.54 | 2.22 | 1.81 | 1.86 | 0.68 | 0.96 | 0.80 | 0.83 | 0.03 | 3.13 | -0.13 | -13.79 |
| Germany | 0.15 | 0.13 | 0.14 | 0.14 | 5.22 | 4.74 | 4.93 | 4.93 | 0.76 | 0.63 | 0.67 | 0.67 | 0.00 | 0.00 | 0.04 | 6.23 |
| Sweden | 0.20 | 0.20 | 0.20 | 0.20 | 3.82 | 4.40 | 4.41 | 4.41 | 0.74 | 0.86 | 0.86 | 0.86 | 0.00 | 0.00 | 0.00 | 0.23 |
| United Kingdom | 0.12 | 0.18 | 0.12 | 0.13 | 5.14 | 5.48 | 5.50 | 5.50 | 0.63 | 0.96 | 0.66 | 0.72 | 0.06 | 8.33 | -0.25 | -25.83 |
| France | 0.08 | 0.09 | 0.10 | 0.10 | 4.83 | 4.62 | 4.64 | 4.64 | 0.40 | 0.43 | 0.45 | 0.45 | 0.00 | 0.00 | 0.02 | 4.65 |
| Italy | 0.12 | 0.09 | 0.08 | 0.08 | 2.43 | 2.52 | 2.40 | 2.40 | 0.29 | 0.23 | 0.20 | 0.20 | 0.00 | 0.00 | -0.03 | -13.22 |
| Denmark | 0.05 | 0.06 | 0.06 | 0.06 | 4.94 | 4.86 | 4.67 | 4.67 | 0.25 | 0.27 | 0.28 | 0.28 | 0.00 | 0.00 | 0.01 | 2.94 |
| Romania | 0.19 | 0.18 | 0.19 | 0.19 | 1.75 | 1.97 | 1.95 | 1.95 | 0.34 | 0.36 | 0.36 | 0.36 | 0.00 | 0.00 | 0.01 | 1.41 |
| Czech Republic | 0.05 | 0.04 | 0.05 | 0.04 | 3.37 | 3.16 | 3.30 | 3.60 | 0.17 | 0.14 | 0.17 | 0.15 | -0.01 | -8.48 | 0.01 | 8.63 |
| Hungary | 0.05 | 0.05 | 0.05 | 0.05 | 2.64 | 2.67 | 2.64 | 2.64 | 0.14 | 0.14 | 0.14 | 0.14 | 0.00 | 0.00 | 0.00 | 2.94 |
| Austria | 0.03 | 0.02 | 0.02 | 0.02 | 3.72 | 3.78 | 3.91 | 3.91 | 0.09 | 0.09 | 0.09 | 0.09 | 0.00 | 0.00 | 0.00 | 3.45 |
| Ireland | 0.02 | 0.03 | 0.03 | 0.03 | 6.54 | 7.31 | 7.20 | 7.20 | 0.16 | 0.19 | 0.18 | 0.18 | 0.00 | 0.00 | -0.01 | -5.26 |
| Lithuania | 0.07 | 0.07 | 0.07 | 0.07 | 2.31 | 2.23 | 2.25 | 2.25 | 0.16 | 0.16 | 0.16 | 0.16 | 0.00 | 0.00 | 0.00 | -0.61 |
| **Australia** | 0.73 | 0.74 | 0.73 | 0.73 | 1.54 | 1.78 | 1.58 | 1.58 | 1.12 | 1.33 | 1.15 | 1.15 | 0.00 | 0.00 | -0.18 | -13.27 |
| **Ukraine** | 0.30 | 0.24 | 0.25 | 0.25 | 2.09 | 1.94 | 2.00 | 2.20 | 0.63 | 0.47 | 0.50 | 0.55 | 0.05 | 10.00 | 0.08 | 17.77 |
| **China** | 0.20 | 0.20 | 0.20 | 0.20 | 3.00 | 2.90 | 2.90 | 3.00 | 0.60 | 0.58 | 0.58 | 0.60 | 0.02 | 3.45 | 0.02 | 3.45 |
| **Belarus** | 0.13 | 0.13 | 0.16 | 0.14 | 3.22 | 2.65 | 2.50 | 3.21 | 0.42 | 0.35 | 0.40 | 0.45 | 0.05 | 12.50 | 0.10 | 27.84 |
| **Brazil** | 0.17 | 0.17 | 0.17 | 0.17 | 2.14 | 2.24 | 2.24 | 2.24 | 0.36 | 0.38 | 0.38 | 0.38 | 0.00 | 0.00 | 0.00 | 0.00 |
| **Argentina** | 0.26 | 0.22 | 0.23 | 0.23 | 1.91 | 2.07 | 2.00 | 2.00 | 0.50 | 0.45 | 0.46 | 0.46 | 0.00 | 0.00 | 0.02 | 3.37 |
| **Chile** | 0.13 | 0.11 | 0.13 | 0.13 | 5.35 | 5.41 | 5.60 | 5.60 | 0.68 | 0.60 | 0.70 | 0.70 | 0.00 | 0.00 | 0.11 | 17.65 |
| **Norway** | 0.07 | 0.07 | 0.07 | 0.07 | 3.42 | 3.42 | 3.42 | 3.42 | 0.24 | 0.24 | 0.24 | 0.24 | 0.00 | 0.00 | 0.00 | 0.00 |
| **Turkey** | 0.09 | 0.09 | 0.09 | 0.09 | 2.36 | 2.33 | 2.33 | 2.33 | 0.21 | 0.21 | 0.21 | 0.21 | 0.00 | 0.00 | 0.00 | 0.00 |
| **Kazakhstan** | 0.15 | 0.22 | 0.20 | 0.20 | 1.33 | 1.39 | 1.25 | 1.25 | 0.20 | 0.31 | 0.25 | 0.25 | 0.00 | 0.00 | -0.06 | -18.03 |
| **Serbia** | 0.03 | 0.03 | 0.03 | 0.03 | 2.20 | 2.20 | 2.20 | 2.20 | 0.07 | 0.07 | 0.07 | 0.07 | 0.00 | 0.00 | 0.00 | 0.00 |
| **Others** | 0.28 | 0.28 | 0.28 | 0.28 | 1.61 | 1.63 | 1.63 | 1.63 | 0.44 | 0.46 | 0.46 | 0.46 | 0.00 | 0.00 | 0.00 | 0.22 |

World and Selected Countries and Regions

**Table 07 Rye Area, Yield, and Production**

| Country / Region | Area (Million hectares) | | | | Yield (Metric tons per hectare) | | | | Production (Million metric tons) | | | | Change in Production | | | |
|---|---|---|---|---|---|---|---|---|---|---|---|---|---|---|---|---|
| | | | | | | | | | | | | | From last month | | From last year | |
| | 2012/13 | Prel. 2013/14 | 2014/15 Proj. Jul | 2014/15 Proj. Aug | 2012/13 | Prel. 2013/14 | 2014/15 Proj. Jul | 2014/15 Proj. Aug | 2012/13 | Prel. 2013/14 | 2014/15 Proj. Jul | 2014/15 Proj. Aug | MMT | Percent | MMT | Percent |
| **World** | 5.03 | 5.44 | 5.69 | 5.38 | 2.74 | 2.91 | 2.69 | 2.75 | 13.78 | 15.82 | 15.30 | 14.78 | -0.52 | -3.39 | -1.04 | -6.54 |
| **United States** | 0.10 | 0.11 | 0.12 | 0.12 | 1.76 | 1.73 | 1.72 | 1.72 | 0.18 | 0.20 | 0.21 | 0.21 | 0.00 | 0.00 | 0.02 | 9.23 |
| **Total Foreign** | 4.93 | 5.32 | 5.56 | 5.26 | 2.76 | 2.93 | 2.71 | 2.77 | 13.60 | 15.62 | 15.09 | 14.57 | -0.52 | -3.43 | -1.05 | -6.74 |
| **European Union - 28** | 2.37 | 2.56 | 2.40 | 2.38 | 3.69 | 3.97 | 3.80 | 3.81 | 8.76 | 10.19 | 9.11 | 9.07 | -0.04 | -0.47 | -1.12 | -10.96 |
| Poland | 1.04 | 1.16 | 1.15 | 1.15 | 2.77 | 2.85 | 2.81 | 2.81 | 2.89 | 3.30 | 3.22 | 3.22 | 0.00 | 0.00 | -0.08 | -2.42 |
| Germany | 0.71 | 0.79 | 0.63 | 0.63 | 5.47 | 5.96 | 5.97 | 5.97 | 3.88 | 4.68 | 3.76 | 3.76 | 0.00 | 0.00 | -0.92 | -19.69 |
| Spain | 0.16 | 0.16 | 0.16 | 0.16 | 1.60 | 2.47 | 1.75 | 1.89 | 0.26 | 0.38 | 0.28 | 0.30 | 0.02 | 7.14 | -0.08 | -21.67 |
| Lithuania | 0.06 | 0.05 | 0.05 | 0.05 | 2.80 | 1.96 | 2.40 | 2.40 | 0.16 | 0.10 | 0.12 | 0.12 | 0.00 | 0.00 | 0.02 | 25.00 |
| Latvia | 0.04 | 0.03 | 0.03 | 0.03 | 3.35 | 2.62 | 2.93 | 2.93 | 0.12 | 0.08 | 0.09 | 0.09 | 0.00 | 0.00 | 0.01 | 15.79 |
| France | 0.03 | 0.03 | 0.03 | 0.03 | 5.00 | 4.90 | 5.00 | 5.00 | 0.16 | 0.14 | 0.15 | 0.15 | 0.00 | 0.00 | 0.01 | 5.63 |
| Denmark | 0.07 | 0.09 | 0.09 | 0.09 | 5.91 | 6.14 | 6.24 | 6.24 | 0.38 | 0.53 | 0.53 | 0.53 | 0.00 | 0.00 | 0.00 | -0.75 |
| Czech Republic | 0.03 | 0.04 | 0.04 | 0.03 | 4.90 | 4.76 | 4.43 | 4.64 | 0.15 | 0.18 | 0.18 | 0.12 | -0.06 | -34.46 | -0.06 | -34.09 |
| Austria | 0.05 | 0.06 | 0.06 | 0.06 | 4.18 | 4.20 | 4.45 | 4.45 | 0.21 | 0.24 | 0.25 | 0.25 | 0.00 | 0.00 | 0.01 | 4.26 |
| Sweden | 0.02 | 0.02 | 0.02 | 0.02 | 6.36 | 7.45 | 7.50 | 7.50 | 0.14 | 0.15 | 0.15 | 0.15 | 0.00 | 0.00 | 0.00 | 0.67 |
| Hungary | 0.04 | 0.04 | 0.04 | 0.04 | 2.23 | 3.00 | 2.86 | 2.86 | 0.08 | 0.11 | 0.10 | 0.10 | 0.00 | 0.00 | -0.01 | -4.76 |
| Slovakia | 0.02 | 0.02 | 0.02 | 0.02 | 3.06 | 3.95 | 3.20 | 3.80 | 0.05 | 0.09 | 0.06 | 0.06 | -0.01 | -10.94 | -0.03 | -34.48 |
| Finland | 0.02 | 0.01 | 0.01 | 0.01 | 3.10 | 2.17 | 2.50 | 2.50 | 0.07 | 0.03 | 0.03 | 0.03 | 0.00 | 0.00 | 0.00 | 15.38 |
| United Kingdom | 0.01 | 0.01 | 0.01 | 0.01 | 7.00 | 7.00 | 7.00 | 7.00 | 0.04 | 0.04 | 0.04 | 0.04 | 0.00 | 0.00 | 0.00 | 0.00 |
| Romania | 0.01 | 0.01 | 0.01 | 0.01 | 2.00 | 2.75 | 3.00 | 3.00 | 0.02 | 0.02 | 0.02 | 0.02 | 0.00 | 0.00 | 0.00 | 9.09 |
| Greece | 0.02 | 0.02 | 0.02 | 0.02 | 1.83 | 1.83 | 1.83 | 1.83 | 0.03 | 0.03 | 0.03 | 0.03 | 0.00 | 0.00 | 0.00 | 0.00 |
| Portugal | 0.02 | 0.02 | 0.02 | 0.02 | 0.77 | 0.86 | 0.86 | 0.86 | 0.02 | 0.02 | 0.02 | 0.02 | 0.00 | 0.00 | 0.00 | 0.00 |
| **Russia** | 1.42 | 1.78 | 2.00 | 2.00 | 1.50 | 1.89 | 1.75 | 1.75 | 2.13 | 3.36 | 3.50 | 3.50 | 0.00 | 0.00 | 0.14 | 4.17 |
| **Belarus** | 0.39 | 0.32 | 0.60 | 0.33 | 2.76 | 2.01 | 2.17 | 2.42 | 1.08 | 0.65 | 1.30 | 0.80 | -0.50 | -38.46 | 0.15 | 23.46 |
| **Ukraine** | 0.30 | 0.28 | 0.20 | 0.19 | 2.27 | 2.29 | 2.00 | 2.30 | 0.68 | 0.64 | 0.40 | 0.43 | 0.03 | 6.25 | -0.21 | -33.39 |
| **Canada** | 0.12 | 0.09 | 0.08 | 0.08 | 2.74 | 2.47 | 2.53 | 2.53 | 0.34 | 0.21 | 0.19 | 0.19 | 0.00 | 0.00 | -0.02 | -9.52 |
| **Turkey** | 0.14 | 0.14 | 0.14 | 0.14 | 2.59 | 2.50 | 2.50 | 2.50 | 0.37 | 0.35 | 0.35 | 0.35 | 0.00 | 0.00 | 0.00 | 0.00 |
| **Argentina** | 0.02 | 0.04 | 0.03 | 0.03 | 1.74 | 1.49 | 1.60 | 1.60 | 0.04 | 0.05 | 0.04 | 0.04 | 0.00 | 0.00 | -0.01 | -23.08 |
| **Kazakhstan** | 0.05 | 0.04 | 0.04 | 0.04 | 1.00 | 1.10 | 1.25 | 1.25 | 0.05 | 0.04 | 0.05 | 0.05 | 0.00 | 0.00 | 0.01 | 16.28 |
| **Australia** | 0.06 | 0.04 | 0.04 | 0.04 | 0.70 | 0.57 | 0.57 | 0.57 | 0.04 | 0.02 | 0.02 | 0.02 | 0.00 | 0.00 | 0.00 | 0.00 |
| **Switzerland** | 0.00 | 0.00 | 0.00 | 0.00 | 5.50 | 5.50 | 5.50 | 5.50 | 0.01 | 0.01 | 0.01 | 0.01 | 0.00 | 0.00 | 0.00 | 0.00 |
| **Others** | 0.05 | 0.05 | 0.05 | 0.05 | 2.26 | 2.17 | 2.33 | 2.33 | 0.10 | 0.10 | 0.11 | 0.11 | 0.00 | 0.00 | 0.01 | 9.80 |

World and Selected Countries and Regions

**Table 08 Sorghum Area, Yield, and Production**

| Country / Region | Area (Million hectares) | | | | Yield (Metric tons per hectare) | | | | Production (Million metric tons) | | | | Change in Production | | | |
|---|---|---|---|---|---|---|---|---|---|---|---|---|---|---|---|---|
| | | | 2014/15 Proj. | | | | 2014/15 Proj. | | | | 2014/15 Proj. | | From last month | | From last year | |
| | 2012/13 | Prel. 2013/14 | Jul | Aug | 2012/13 | Prel. 2013/14 | Jul | Aug | 2012/13 | Prel. 2013/14 | Jul | Aug | MMT | Percent | MMT | Percent |
| **World** | 38.16 | 40.52 | 40.95 | 40.21 | 1.52 | 1.47 | 1.57 | 1.59 | 57.93 | 59.69 | 64.08 | 63.81 | -0.27 | -0.43 | 4.12 | 6.91 |
| **United States** | 2.01 | 2.64 | 2.59 | 2.59 | 3.13 | 3.74 | 4.02 | 4.21 | 6.27 | 9.88 | 10.41 | 10.91 | 0.49 | 4.71 | 1.02 | 10.35 |
| **Total Foreign** | 36.15 | 37.88 | 38.36 | 37.62 | 1.43 | 1.31 | 1.40 | 1.41 | 51.66 | 49.81 | 53.67 | 52.91 | -0.76 | -1.42 | 3.10 | 6.22 |
| **Sub-Saharan Africa** | | | | | | | | | | | | | | | | |
| Nigeria | 4.77 | 5.00 | 5.00 | 5.00 | 1.25 | 1.30 | 1.30 | 1.30 | 5.94 | 6.50 | 6.50 | 6.50 | 0.00 | 0.00 | 0.00 | 0.00 |
| Sudan | 4.10 | 5.60 | 5.60 | 5.60 | 1.10 | 0.40 | 0.68 | 0.68 | 4.52 | 2.25 | 3.80 | 3.80 | 0.00 | 0.00 | 1.55 | 68.96 |
| Ethiopia | 1.71 | 1.80 | 1.80 | 1.80 | 2.11 | 2.22 | 2.22 | 2.22 | 3.60 | 4.00 | 4.00 | 4.00 | 0.00 | 0.00 | 0.00 | 0.00 |
| Burkina | 1.79 | 1.80 | 1.80 | 1.80 | 1.08 | 1.08 | 1.06 | 1.06 | 1.92 | 1.94 | 1.90 | 1.90 | 0.00 | 0.00 | -0.04 | -2.06 |
| Tanzania | 0.84 | 0.85 | 0.85 | 0.85 | 1.00 | 0.94 | 0.94 | 0.94 | 0.84 | 0.80 | 0.80 | 0.80 | 0.00 | 0.00 | 0.00 | 0.00 |
| Niger | 3.11 | 3.00 | 3.00 | 3.00 | 0.44 | 0.43 | 0.40 | 0.40 | 1.38 | 1.29 | 1.20 | 1.20 | 0.00 | 0.00 | -0.09 | -6.76 |
| Uganda | 0.37 | 0.35 | 0.35 | 0.35 | 0.90 | 0.91 | 0.91 | 0.91 | 0.34 | 0.32 | 0.32 | 0.32 | 0.00 | 0.00 | 0.00 | 0.00 |
| Mozambique | 0.62 | 0.62 | 0.62 | 0.62 | 0.39 | 0.30 | 0.48 | 0.48 | 0.24 | 0.19 | 0.30 | 0.30 | 0.00 | 0.00 | 0.11 | 59.57 |
| Ghana | 0.23 | 0.25 | 0.25 | 0.25 | 1.21 | 1.11 | 1.20 | 1.20 | 0.28 | 0.28 | 0.30 | 0.30 | 0.00 | 0.00 | 0.02 | 8.30 |
| South Africa | 0.06 | 0.08 | 0.08 | 0.08 | 2.33 | 3.24 | 2.81 | 2.81 | 0.15 | 0.26 | 0.23 | 0.23 | 0.00 | 0.00 | -0.03 | -12.11 |
| **South Asia** | | | | | | | | | | | | | | | | |
| India | 6.30 | 5.90 | 6.20 | 5.50 | 0.84 | 0.89 | 0.97 | 0.96 | 5.30 | 5.25 | 6.00 | 5.30 | -0.70 | -11.67 | 0.05 | 0.95 |
| Pakistan | 0.24 | 0.24 | 0.24 | 0.24 | 0.60 | 0.60 | 0.60 | 0.60 | 0.15 | 0.15 | 0.15 | 0.15 | 0.00 | 0.00 | 0.00 | 0.00 |
| **South America** | | | | | | | | | | | | | | | | |
| Argentina | 1.05 | 1.00 | 1.10 | 1.10 | 4.48 | 4.20 | 4.36 | 4.36 | 4.70 | 4.20 | 4.80 | 4.80 | 0.00 | 0.00 | 0.60 | 14.29 |
| Brazil | 0.80 | 0.85 | 0.90 | 0.90 | 2.62 | 2.82 | 2.67 | 2.67 | 2.10 | 2.40 | 2.40 | 2.40 | 0.00 | 0.00 | 0.00 | 0.00 |
| Mexico | 1.64 | 1.98 | 1.83 | 1.83 | 3.76 | 3.62 | 3.83 | 3.83 | 6.17 | 7.17 | 7.00 | 7.00 | 0.00 | 0.00 | -0.17 | -2.37 |
| China | 0.62 | 0.65 | 0.67 | 0.67 | 4.10 | 4.15 | 4.18 | 4.18 | 2.56 | 2.70 | 2.80 | 2.80 | 0.00 | 0.00 | 0.10 | 3.70 |
| Australia | 0.65 | 0.49 | 0.67 | 0.67 | 3.44 | 2.25 | 3.06 | 3.06 | 2.23 | 1.11 | 2.05 | 2.05 | 0.00 | 0.00 | 0.94 | 85.19 |
| Egypt | 0.14 | 0.14 | 0.14 | 0.14 | 5.29 | 5.29 | 5.29 | 5.29 | 0.76 | 0.76 | 0.76 | 0.76 | 0.00 | 0.00 | 0.00 | 0.00 |
| **European Union - 28** | 0.11 | 0.12 | 0.11 | 0.11 | 4.34 | 4.97 | 5.05 | 5.05 | 0.50 | 0.60 | 0.57 | 0.58 | 0.01 | 1.05 | -0.02 | -3.36 |
| France | 0.04 | 0.05 | 0.05 | 0.05 | 5.71 | 5.10 | 5.56 | 5.56 | 0.24 | 0.26 | 0.25 | 0.25 | 0.00 | 0.00 | -0.01 | -3.85 |
| Italy | 0.04 | 0.04 | 0.04 | 0.04 | 4.27 | 6.22 | 6.22 | 6.22 | 0.16 | 0.23 | 0.23 | 0.23 | 0.00 | 0.00 | 0.00 | 0.00 |
| **Others** | 6.98 | 7.15 | 7.14 | 7.10 | 1.15 | 1.07 | 1.09 | 1.09 | 7.99 | 7.66 | 7.80 | 7.73 | -0.07 | -0.90 | 0.07 | 0.90 |

World and Selected Countries and Regions

# Table 09 Rice Area, Yield, and Production

## World and Selected Countries and Regions

| Country / Region | Area (Million hectares) | | | | Yield (Metric tons per hectare) | | | | Production (Million metric tons) | | | | Change in Production | | | |
|---|---|---|---|---|---|---|---|---|---|---|---|---|---|---|---|---|
| | | | 2014/15 Proj. | | | | 2014/15 Proj. | | | | 2014/15 Proj. | | From last month | | From last year | |
| | 2012/13 | Prel. 2013/14 | Jul | Aug | 2012/13 | Prel. 2013/14 | Jul | Aug | 2012/13 | Prel. 2013/14 | Jul | Aug | MMT | Percent | MMT | Percent |
| **World** | 158.01 | 160.96 | 161.48 | 161.04 | 4.45 | 4.41 | 4.43 | 4.42 | 471.70 | 475.80 | 479.43 | 477.35 | -2.08 | -0.43 | 1.55 | 0.33 |
| **United States** | 1.08 | 1.00 | 1.23 | 1.23 | 8.35 | 8.62 | 8.37 | 8.47 | 6.34 | 6.12 | 7.23 | 7.32 | 0.09 | 1.23 | 1.20 | 19.64 |
| **Total Foreign** | 156.93 | 159.97 | 160.26 | 159.81 | 4.42 | 4.38 | 4.40 | 4.39 | 465.37 | 469.69 | 472.21 | 470.03 | -2.17 | -0.46 | 0.35 | 0.07 |
| **East Asia** | | | | | | | | | | | | | | | | |
| China | 30.14 | 30.31 | 30.60 | 30.60 | 6.78 | 6.72 | 6.72 | 6.72 | 143.00 | 142.53 | 144.00 | 144.00 | 0.00 | 0.00 | 1.47 | 1.03 |
| Japan | 1.58 | 1.60 | 1.59 | 1.59 | 6.74 | 6.73 | 6.65 | 6.65 | 7.83 | 7.70 | 7.70 | 7.70 | 0.00 | 0.00 | -0.13 | -1.69 |
| Korea, South | 0.85 | 0.83 | 0.83 | 0.83 | 6.37 | 6.76 | 6.70 | 6.70 | 4.01 | 4.23 | 4.15 | 4.15 | 0.00 | 0.00 | -0.08 | -1.89 |
| Korea, North | 0.58 | 0.57 | 0.57 | 0.57 | 4.62 | 5.07 | 4.90 | 4.90 | 1.74 | 1.88 | 1.80 | 1.80 | 0.00 | 0.00 | -0.08 | -4.26 |
| **South Asia** | | | | | | | | | | | | | | | | |
| India | 42.41 | 43.94 | 43.80 | 43.40 | 3.72 | 3.63 | 3.56 | 3.56 | 105.24 | 106.29 | 104.00 | 103.00 | -1.00 | -0.96 | -3.29 | -3.10 |
| Bangladesh | 11.65 | 11.77 | 11.82 | 11.80 | 4.35 | 4.38 | 4.42 | 4.40 | 33.82 | 34.39 | 34.80 | 34.60 | -0.20 | -0.57 | 0.21 | 0.61 |
| Pakistan | 2.40 | 2.76 | 2.76 | 2.76 | 3.63 | 3.59 | 3.64 | 3.64 | 5.80 | 6.60 | 6.70 | 6.70 | 0.00 | 0.00 | 0.10 | 1.52 |
| **Southeast Asia** | | | | | | | | | | | | | | | | |
| Indonesia | 12.19 | 12.00 | 12.16 | 12.16 | 4.72 | 4.72 | 4.88 | 4.79 | 36.55 | 36.00 | 37.70 | 37.00 | -0.70 | -1.86 | 1.00 | 2.78 |
| Vietnam | 7.86 | 7.78 | 7.78 | 7.78 | 5.60 | 5.76 | 5.80 | 5.80 | 27.54 | 28.00 | 28.20 | 28.20 | 0.00 | 0.00 | 0.20 | 0.71 |
| Thailand | 10.84 | 10.92 | 10.90 | 10.90 | 2.82 | 2.84 | 2.85 | 2.85 | 20.20 | 20.46 | 20.50 | 20.50 | 0.00 | 0.00 | 0.04 | 0.20 |
| Burma | 7.04 | 7.05 | 7.07 | 7.07 | 2.60 | 2.65 | 2.69 | 2.69 | 11.72 | 11.96 | 12.15 | 12.15 | 0.00 | 0.00 | 0.19 | 1.61 |
| Philippines | 4.70 | 4.80 | 4.89 | 4.89 | 3.86 | 3.90 | 3.96 | 3.96 | 11.43 | 11.81 | 12.20 | 12.20 | 0.00 | 0.00 | 0.39 | 3.28 |
| Cambodia | 2.98 | 2.97 | 3.05 | 3.05 | 2.45 | 2.49 | 2.51 | 2.51 | 4.67 | 4.73 | 4.90 | 4.90 | 0.00 | 0.00 | 0.18 | 3.70 |
| Laos | 0.93 | 0.87 | 0.90 | 0.90 | 2.81 | 2.68 | 2.73 | 2.73 | 1.66 | 1.47 | 1.55 | 1.55 | 0.00 | 0.00 | 0.09 | 5.80 |
| Malaysia | 0.69 | 0.69 | 0.69 | 0.69 | 3.79 | 3.91 | 4.02 | 4.02 | 1.69 | 1.76 | 1.80 | 1.80 | 0.00 | 0.00 | 0.05 | 2.56 |
| **South America** | | | | | | | | | | | | | | | | |
| Brazil | 2.39 | 2.40 | 2.45 | 2.45 | 4.95 | 5.09 | 5.10 | 5.01 | 8.04 | 8.30 | 8.50 | 8.35 | -0.15 | -1.76 | 0.05 | 0.60 |
| Peru | 0.39 | 0.41 | 0.40 | 0.40 | 7.72 | 7.72 | 7.61 | 7.61 | 2.10 | 2.16 | 2.10 | 2.10 | 0.00 | 0.00 | -0.06 | -2.60 |
| **Sub-Saharan Africa** | | | | | | | | | | | | | | | | |
| Nigeria | 2.00 | 2.50 | 2.30 | 2.30 | 1.88 | 1.76 | 1.76 | 1.76 | 2.37 | 2.77 | 2.55 | 2.55 | 0.00 | 0.00 | -0.22 | -8.01 |
| Madagascar | 1.35 | 1.45 | 1.45 | 1.45 | 3.37 | 2.49 | 3.10 | 3.10 | 2.91 | 2.31 | 2.88 | 2.88 | 0.00 | 0.00 | 0.57 | 24.62 |
| **European Union - 28** | 0.45 | 0.44 | 0.43 | 0.44 | 6.68 | 6.39 | 6.52 | 6.49 | 2.09 | 1.94 | 1.96 | 1.97 | 0.01 | 0.46 | 0.03 | 1.44 |
| Italy | 0.24 | 0.22 | 0.22 | 0.22 | 6.60 | 6.36 | 6.36 | 6.30 | 1.09 | 0.96 | 0.96 | 0.97 | 0.00 | 0.42 | 0.00 | 0.42 |
| Spain | 0.11 | 0.11 | 0.11 | 0.11 | 7.73 | 7.53 | 7.79 | 7.79 | 0.62 | 0.60 | 0.62 | 0.62 | 0.00 | 0.00 | 0.02 | 3.36 |
| **Egypt** | 0.77 | 0.79 | 0.80 | 0.80 | 8.80 | 8.95 | 8.93 | 8.93 | 4.68 | 4.88 | 4.90 | 4.90 | 0.00 | 0.00 | 0.02 | 0.41 |
| **Iran** | 0.53 | 0.59 | 0.60 | 0.60 | 4.43 | 4.24 | 4.25 | 4.25 | 1.54 | 1.65 | 1.68 | 1.68 | 0.00 | 0.00 | 0.03 | 2.00 |
| **Others** | 12.21 | 12.53 | 12.43 | 12.41 | 3.07 | 3.11 | 3.09 | 3.09 | 24.84 | 25.75 | 25.48 | 25.35 | -0.13 | -0.52 | -0.40 | -1.55 |

Yield is on a rough basis, before the milling process
Production is on a milled basis, after the milling process

# Table 10 Total Oilseed Area, Yield, and Production

## World and Selected Countries and Regions

| Country / Region | Area (Million hectares) | | | | Yield (Metric tons per hectare) | | | | Production (Million metric tons) | | | | Change in Production | | | |
|---|---|---|---|---|---|---|---|---|---|---|---|---|---|---|---|---|
| | 2012/13 | Prel. 2013/14 | 2014/15 Proj. Jul | Aug | 2012/13 | Prel. 2013/14 | 2014/15 Proj. Jul | Aug | 2012/13 | Prel. 2013/14 | 2014/15 Proj. Jul | Aug | From last month MMT | Percent | From last year MMT | Percent |
| **World Total** | -- | -- | -- | -- | -- | -- | -- | -- | 474.53 | 503.91 | 521.86 | 521.80 | -0.06 | -0.01 | 17.90 | 3.55 |
| **Total Foreign** | -- | -- | -- | -- | -- | -- | -- | -- | 381.39 | 406.77 | 408.81 | 408.13 | -0.68 | -0.17 | 1.36 | 0.33 |
| **Oilseed, Copra** | -- | -- | -- | -- | -- | -- | -- | -- | 5.80 | 5.58 | 5.53 | 5.53 | 0.00 | 0.00 | -0.05 | -0.88 |
| **Oilseed, Palm Kernel** | -- | -- | -- | -- | -- | -- | -- | -- | 14.79 | 15.66 | 16.46 | 16.47 | 0.01 | 0.07 | 0.80 | 5.13 |
| **Major OilSeeds** | 226.88 | 229.25 | 234.60 | 233.81 | 2.00 | 2.11 | 2.13 | 2.14 | 453.94 | 482.67 | 499.88 | 499.81 | -0.07 | -0.01 | 17.14 | 3.55 |
| **United States** | 36.71 | 35.29 | 39.80 | 40.02 | 2.54 | 2.75 | 2.84 | 2.84 | 93.15 | 97.14 | 113.05 | 113.68 | 0.63 | 0.55 | 16.54 | 17.03 |
| **Foreign Oilseeds** | 190.17 | 193.96 | 194.81 | 193.79 | 1.90 | 1.99 | 1.99 | 1.99 | 360.79 | 385.53 | 386.83 | 386.13 | -0.70 | -0.18 | 0.61 | 0.16 |
| **South America** | 56.62 | 59.33 | 60.02 | 59.91 | 2.72 | 2.75 | 2.80 | 2.80 | 153.72 | 163.34 | 168.07 | 167.79 | -0.29 | -0.17 | 4.44 | 2.72 |
| Brazil | 28.77 | 31.26 | 31.88 | 31.78 | 2.94 | 2.91 | 2.96 | 2.96 | 84.67 | 90.96 | 94.47 | 94.19 | -0.28 | -0.29 | 3.23 | 3.55 |
| Argentina | 21.76 | 21.99 | 22.11 | 22.11 | 2.47 | 2.61 | 2.65 | 2.65 | 53.68 | 57.50 | 58.53 | 58.53 | 0.00 | 0.00 | 1.02 | 1.78 |
| Paraguay | 3.34 | 3.26 | 3.29 | 3.28 | 2.53 | 2.55 | 2.57 | 2.58 | 8.44 | 8.32 | 8.46 | 8.45 | -0.01 | -0.12 | 0.13 | 1.59 |
| Bolivia | 1.26 | 1.18 | 1.30 | 1.30 | 2.29 | 2.26 | 2.15 | 2.15 | 2.88 | 2.67 | 2.80 | 2.80 | 0.00 | 0.00 | 0.13 | 4.87 |
| **China** | 25.43 | 24.94 | 24.25 | 24.23 | 2.35 | 2.35 | 2.38 | 2.38 | 59.79 | 58.62 | 57.76 | 57.76 | 0.00 | 0.00 | -0.86 | -1.47 |
| **South Asia** | 39.64 | 41.38 | 40.68 | 40.48 | 1.05 | 1.04 | 1.04 | 1.04 | 41.51 | 42.95 | 42.40 | 42.20 | -0.20 | -0.47 | -0.75 | -1.75 |
| India | 35.35 | 37.18 | 36.53 | 36.33 | 1.02 | 1.01 | 1.02 | 1.02 | 36.10 | 37.57 | 37.10 | 36.90 | -0.20 | -0.54 | -0.67 | -1.78 |
| Pakistan | 3.96 | 3.87 | 3.82 | 3.82 | 1.30 | 1.32 | 1.32 | 1.32 | 5.15 | 5.12 | 5.04 | 5.04 | 0.00 | 0.00 | -0.08 | -1.56 |
| **European Union - 28** | 11.40 | 12.04 | 11.95 | 11.92 | 2.47 | 2.63 | 2.70 | 2.74 | 28.13 | 31.69 | 32.29 | 32.61 | 0.32 | 1.00 | 0.93 | 2.92 |
| France | 2.32 | 2.25 | 2.28 | 2.25 | 3.07 | 2.69 | 3.08 | 3.13 | 7.14 | 6.06 | 7.02 | 7.04 | 0.02 | 0.28 | 0.98 | 16.17 |
| Germany | 1.33 | 1.49 | 1.46 | 1.46 | 3.67 | 3.92 | 4.07 | 4.07 | 4.88 | 5.82 | 5.94 | 5.93 | -0.01 | -0.13 | 0.11 | 1.82 |
| Poland | 0.84 | 0.96 | 0.90 | 0.86 | 2.65 | 2.91 | 2.98 | 3.09 | 2.22 | 2.80 | 2.68 | 2.66 | -0.02 | -0.75 | -0.14 | -4.99 |
| Romania | 1.30 | 1.47 | 1.49 | 1.48 | 1.34 | 2.06 | 1.91 | 2.07 | 1.75 | 3.01 | 2.84 | 3.07 | 0.23 | 8.10 | 0.06 | 1.99 |
| Hungary | 0.81 | 0.84 | 0.89 | 0.89 | 2.18 | 2.48 | 2.30 | 2.28 | 1.77 | 2.08 | 2.04 | 2.02 | -0.02 | -0.74 | -0.06 | -2.69 |
| United Kingdom | 0.76 | 0.72 | 0.71 | 0.72 | 3.38 | 2.97 | 3.58 | 3.57 | 2.56 | 2.13 | 2.53 | 2.57 | 0.04 | 1.62 | 0.44 | 20.83 |
| **Former Soviet Union - 12** | 19.87 | 21.07 | 22.25 | 21.85 | 1.41 | 1.66 | 1.55 | 1.57 | 27.95 | 34.92 | 34.59 | 34.37 | -0.22 | -0.64 | -0.56 | -1.59 |
| Russia | 8.45 | 9.11 | 9.80 | 9.50 | 1.29 | 1.49 | 1.47 | 1.47 | 10.87 | 13.58 | 14.40 | 14.00 | -0.40 | -2.78 | 0.42 | 3.07 |
| Ukraine | 7.46 | 7.95 | 8.30 | 8.20 | 1.70 | 2.10 | 1.87 | 1.91 | 12.71 | 16.73 | 15.50 | 15.70 | 0.20 | 1.29 | -1.03 | -6.13 |
| Uzbekistan | 1.35 | 1.29 | 1.29 | 1.29 | 1.31 | 1.30 | 1.30 | 1.30 | 1.77 | 1.67 | 1.67 | 1.67 | 0.00 | 0.00 | 0.00 | 0.00 |
| **Canada** | 10.52 | 9.86 | 10.15 | 9.85 | 1.81 | 2.36 | 2.15 | 2.17 | 19.04 | 23.25 | 21.88 | 21.43 | -0.45 | -2.06 | -1.82 | -7.84 |
| **Africa** | 16.75 | 16.03 | 16.26 | 16.31 | 0.89 | 0.94 | 0.92 | 0.92 | 14.96 | 15.04 | 15.01 | 15.03 | 0.02 | 0.16 | -0.01 | -0.09 |
| Nigeria | 3.29 | 3.23 | 3.24 | 3.24 | 1.15 | 1.11 | 1.11 | 1.11 | 3.79 | 3.58 | 3.59 | 3.59 | 0.00 | 0.00 | 0.01 | 0.22 |
| **Southeast Asia** | 3.47 | 3.46 | 3.46 | 3.46 | 1.35 | 1.44 | 1.45 | 1.45 | 4.70 | 4.98 | 5.00 | 5.00 | 0.00 | 0.00 | 0.02 | 0.48 |
| Indonesia | 1.14 | 1.12 | 1.09 | 1.09 | 1.54 | 1.61 | 1.63 | 1.63 | 1.76 | 1.79 | 1.78 | 1.78 | 0.00 | 0.00 | -0.01 | -0.56 |
| **Australia** | 3.80 | 3.16 | 3.09 | 3.09 | 1.51 | 1.64 | 1.44 | 1.44 | 5.72 | 5.17 | 4.44 | 4.44 | 0.00 | 0.00 | -0.74 | -14.21 |
| **Others** | 2.68 | 2.69 | 2.69 | 2.71 | 1.97 | 2.07 | 2.00 | 2.04 | 5.28 | 5.56 | 5.40 | 5.51 | 0.11 | 2.11 | -0.05 | -0.90 |

World Total and Total Foreign:: (Major Oilseeds plus copra and palm kernel)

Major Oilseeds:  (soybeans, sunflowerseeds, peanuts(inshell), cottonseed and rapeseed)

# Table 11 Soybean Area, Yield, and Production

| Country / Region | Area (Million hectares) | | | | Yield (Metric tons per hectare) | | | | Production (Million metric tons) | | | | Change in Production | | | |
|---|---|---|---|---|---|---|---|---|---|---|---|---|---|---|---|---|
| | | | | | | | | | | | | | From last month | | From last year | |
| | 2012/13 | Prel. 2013/14 | 2014/15 Proj. Jul | 2014/15 Proj. Aug | 2012/13 | Prel. 2013/14 | 2014/15 Proj. Jul | 2014/15 Proj. Aug | 2012/13 | Prel. 2013/14 | 2014/15 Proj. Jul | 2014/15 Proj. Aug | MMT | Percent | MMT | Percent |
| **World** | 109.44 | 112.85 | 117.15 | 116.58 | 2.45 | 2.52 | 2.60 | 2.61 | 267.86 | 283.95 | 304.79 | 304.69 | -0.10 | -0.03 | 20.75 | 7.31 |
| **United States** | 30.82 | 30.70 | 34.02 | 34.02 | 2.68 | 2.92 | 3.04 | 3.05 | 82.56 | 89.51 | 103.42 | 103.85 | 0.43 | 0.41 | 14.34 | 16.02 |
| **Total Foreign** | 78.61 | 82.15 | 83.13 | 82.56 | 2.36 | 2.37 | 2.42 | 2.43 | 185.30 | 194.44 | 201.37 | 200.85 | -0.52 | -0.26 | 6.41 | 3.30 |
| **South America** | | | | | | | | | | | | | | | | |
| Brazil | 27.70 | 29.90 | 30.50 | 30.50 | 2.96 | 2.93 | 2.98 | 2.98 | 82.00 | 87.50 | 91.00 | 91.00 | 0.00 | 0.00 | 3.50 | 4.00 |
| Argentina | 19.40 | 19.80 | 19.60 | 19.60 | 2.54 | 2.73 | 2.76 | 2.76 | 49.30 | 54.00 | 54.00 | 54.00 | 0.00 | 0.00 | 0.00 | 0.00 |
| Paraguay | 3.16 | 3.10 | 3.10 | 3.10 | 2.60 | 2.61 | 2.65 | 2.65 | 8.20 | 8.10 | 8.20 | 8.20 | 0.00 | 0.00 | 0.10 | 1.23 |
| Bolivia | 1.09 | 1.00 | 1.10 | 1.10 | 2.42 | 2.40 | 2.27 | 2.27 | 2.63 | 2.40 | 2.50 | 2.50 | 0.00 | 0.00 | 0.10 | 4.17 |
| Uruguay | 1.30 | 1.45 | 1.25 | 1.25 | 2.81 | 2.41 | 2.72 | 2.72 | 3.65 | 3.50 | 3.40 | 3.40 | 0.00 | 0.00 | -0.10 | -2.86 |
| **East Asia** | | | | | | | | | | | | | | | | |
| China | 7.17 | 6.85 | 6.70 | 6.70 | 1.82 | 1.78 | 1.79 | 1.79 | 13.05 | 12.20 | 12.00 | 12.00 | 0.00 | 0.00 | -0.20 | -1.64 |
| Korea, South | 0.08 | 0.08 | 0.07 | 0.07 | 1.52 | 1.93 | 1.72 | 1.72 | 0.12 | 0.15 | 0.13 | 0.13 | 0.00 | 0.00 | -0.03 | -17.53 |
| Korea, North | 0.12 | 0.12 | 0.12 | 0.12 | 1.46 | 1.41 | 1.40 | 1.40 | 0.17 | 0.16 | 0.17 | 0.17 | 0.00 | 0.00 | 0.00 | 1.23 |
| Japan | 0.13 | 0.13 | 0.13 | 0.13 | 1.80 | 1.56 | 1.64 | 1.64 | 0.24 | 0.20 | 0.21 | 0.21 | 0.00 | 0.00 | 0.01 | 3.54 |
| **India** | 10.80 | 12.20 | 11.60 | 11.00 | 1.06 | 0.90 | 1.00 | 1.00 | 11.50 | 11.00 | 11.60 | 11.00 | -0.60 | -5.17 | 0.00 | 0.00 |
| Canada | 1.68 | 1.82 | 2.12 | 2.12 | 3.03 | 2.86 | 2.89 | 2.89 | 5.09 | 5.20 | 6.12 | 6.12 | 0.00 | 0.00 | 0.92 | 17.69 |
| **Former Soviet Union - 12** | | | | | | | | | | | | | | | | |
| Russia | 1.35 | 1.20 | 1.80 | 1.80 | 1.39 | 1.36 | 1.50 | 1.50 | 1.88 | 1.64 | 2.70 | 2.70 | 0.00 | 0.00 | 1.06 | 65.04 |
| Ukraine | 1.41 | 1.35 | 1.75 | 1.75 | 1.71 | 2.05 | 2.00 | 2.00 | 2.41 | 2.77 | 3.50 | 3.50 | 0.00 | 0.00 | 0.73 | 26.17 |
| **Southeast Asia** | | | | | | | | | | | | | | | | |
| Indonesia | 0.45 | 0.45 | 0.45 | 0.45 | 1.33 | 1.38 | 1.38 | 1.38 | 0.60 | 0.62 | 0.62 | 0.62 | 0.00 | 0.00 | 0.00 | 0.00 |
| Vietnam | 0.12 | 0.12 | 0.13 | 0.13 | 1.42 | 1.47 | 1.46 | 1.46 | 0.17 | 0.18 | 0.19 | 0.19 | 0.00 | 0.00 | 0.01 | 7.95 |
| Thailand | 0.05 | 0.04 | 0.04 | 0.04 | 1.70 | 1.67 | 1.75 | 1.75 | 0.09 | 0.07 | 0.07 | 0.07 | 0.00 | 0.00 | 0.00 | 0.00 |
| Burma | 0.17 | 0.17 | 0.17 | 0.17 | 1.23 | 1.47 | 1.47 | 1.47 | 0.21 | 0.25 | 0.25 | 0.25 | 0.00 | 0.00 | 0.00 | 0.00 |
| **European Union - 28** | 0.42 | 0.47 | 0.53 | 0.56 | 2.25 | 2.62 | 2.69 | 2.64 | 0.95 | 1.23 | 1.43 | 1.47 | 0.05 | 3.30 | 0.24 | 19.77 |
| Italy | 0.15 | 0.18 | 0.22 | 0.22 | 2.76 | 3.54 | 3.50 | 3.50 | 0.42 | 0.62 | 0.76 | 0.76 | 0.00 | 0.00 | 0.14 | 22.58 |
| France | 0.04 | 0.04 | 0.04 | 0.07 | 2.81 | 2.62 | 2.86 | 2.00 | 0.10 | 0.11 | 0.12 | 0.14 | 0.02 | 16.67 | 0.03 | 27.27 |
| **Sub-Saharan Africa** | | | | | | | | | | | | | | | | |
| South Africa | 0.52 | 0.50 | 0.56 | 0.56 | 1.52 | 1.88 | 1.64 | 1.64 | 0.79 | 0.94 | 0.92 | 0.92 | 0.00 | 0.00 | -0.02 | -2.54 |
| Nigeria | 0.57 | 0.44 | 0.44 | 0.44 | 1.02 | 1.02 | 1.02 | 1.02 | 0.58 | 0.45 | 0.45 | 0.45 | 0.00 | 0.00 | 0.00 | 0.00 |
| Uganda | 0.15 | 0.15 | 0.15 | 0.15 | 1.00 | 1.11 | 1.11 | 1.11 | 0.15 | 0.17 | 0.17 | 0.17 | 0.00 | 0.00 | 0.00 | 0.00 |
| **Serbia** | 0.17 | 0.17 | 0.17 | 0.17 | 1.70 | 2.42 | 2.36 | 2.55 | 0.28 | 0.40 | 0.39 | 0.42 | 0.03 | 7.69 | 0.02 | 5.00 |
| **Mexico** | 0.14 | 0.16 | 0.17 | 0.17 | 1.75 | 1.60 | 1.76 | 1.76 | 0.25 | 0.25 | 0.29 | 0.29 | 0.00 | 0.00 | 0.04 | 14.17 |
| **Iran** | 0.08 | 0.08 | 0.08 | 0.08 | 2.50 | 2.50 | 2.44 | 2.44 | 0.20 | 0.20 | 0.20 | 0.20 | 0.00 | 0.00 | -0.01 | -2.50 |
| **Others** | 0.39 | 0.40 | 0.42 | 0.42 | 2.06 | 2.13 | 2.13 | 2.13 | 0.81 | 0.86 | 0.89 | 0.89 | 0.00 | 0.00 | -0.03 | 3.86 |

World and Selected Countries and Regions

**Table 12 Cottonseed Area, Yield, and Production**

| Country / Region | Area (Million hectares) | | | | Yield (Metric tons per hectare) | | | | Production (Million metric tons) | | | | Change in Production | | | |
|---|---|---|---|---|---|---|---|---|---|---|---|---|---|---|---|---|
| | | | 2014/15 Proj. | | | | 2014/15 Proj. | | | | 2014/15 Proj. | | From last month | | From last year | |
| | 2012/13 | Prel. 2013/14 | Jul | Aug | 2012/13 | Prel. 2013/14 | Jul | Aug | 2012/13 | Prel. 2013/14 | Jul | Aug | MMT | Percent | MMT | Percent |
| **World** | 33.43 | 31.86 | 32.47 | 33.03 | 1.39 | 1.41 | 1.35 | 1.34 | 46.31 | 44.87 | 43.86 | 44.23 | 0.37 | 0.85 | -0.64 | -1.42 |
| **United States** | 3.79 | 3.05 | 3.93 | 4.14 | 1.36 | 1.25 | 1.28 | 1.26 | 5.14 | 3.81 | 5.02 | 5.24 | 0.22 | 4.38 | 1.43 | 37.45 |
| **Total Foreign** | 29.64 | 28.80 | 28.54 | 28.88 | 1.39 | 1.43 | 1.36 | 1.35 | 41.17 | 41.05 | 38.84 | 38.99 | 0.15 | 0.39 | -2.06 | -5.03 |
| **China** | 5.30 | 4.90 | 4.35 | 4.35 | 2.59 | 2.56 | 2.66 | 2.66 | 13.72 | 12.54 | 11.56 | 11.56 | 0.00 | 0.00 | -0.98 | -7.81 |
| **South Asia** | | | | | | | | | | | | | | | | |
| India | 12.00 | 11.70 | 11.80 | 12.20 | 1.01 | 1.11 | 1.01 | 1.01 | 12.10 | 12.95 | 11.90 | 12.30 | 0.40 | 3.36 | -0.65 | -5.02 |
| Pakistan | 3.00 | 3.00 | 3.00 | 3.00 | 1.33 | 1.37 | 1.37 | 1.37 | 4.00 | 4.10 | 4.10 | 4.10 | 0.00 | 0.00 | 0.00 | 0.00 |
| **Former Soviet Union - 12** | | | | | | | | | | | | | | | | |
| Uzbekistan | 1.35 | 1.29 | 1.29 | 1.29 | 1.31 | 1.30 | 1.30 | 1.30 | 1.77 | 1.67 | 1.67 | 1.67 | 0.00 | 0.00 | 0.00 | 0.00 |
| Turkmenistan | 0.60 | 0.58 | 0.58 | 0.58 | 1.04 | 1.03 | 1.03 | 1.03 | 0.63 | 0.59 | 0.59 | 0.59 | 0.00 | 0.00 | 0.00 | 0.00 |
| Tajikistan | 0.20 | 0.19 | 0.19 | 0.18 | 1.08 | 0.92 | 1.03 | 0.96 | 0.22 | 0.18 | 0.20 | 0.18 | -0.02 | -10.26 | 0.00 | 0.00 |
| Kazakhstan | 0.15 | 0.14 | 0.13 | 0.13 | 1.10 | 0.96 | 0.95 | 0.95 | 0.16 | 0.13 | 0.13 | 0.13 | 0.00 | 0.00 | -0.01 | -4.51 |
| **South America** | | | | | | | | | | | | | | | | |
| Brazil | 0.90 | 1.12 | 1.15 | 1.05 | 2.48 | 2.59 | 2.59 | 2.57 | 2.23 | 2.90 | 2.98 | 2.70 | -0.28 | -9.24 | -0.20 | -6.90 |
| Argentina | 0.36 | 0.56 | 0.57 | 0.57 | 0.72 | 0.75 | 0.75 | 0.75 | 0.26 | 0.42 | 0.43 | 0.43 | 0.00 | 0.00 | 0.01 | 1.19 |
| **Middle East** | | | | | | | | | | | | | | | | |
| Turkey | 0.41 | 0.33 | 0.42 | 0.42 | 2.12 | 2.24 | 2.22 | 2.22 | 0.87 | 0.74 | 0.93 | 0.93 | 0.00 | 0.00 | 0.19 | 25.81 |
| Syria | 0.13 | 0.11 | 0.06 | 0.06 | 2.72 | 2.51 | 1.62 | 1.62 | 0.34 | 0.28 | 0.10 | 0.10 | 0.00 | 0.00 | -0.18 | -64.86 |
| Iran | 0.11 | 0.10 | 0.11 | 0.11 | 1.14 | 0.90 | 0.87 | 0.87 | 0.12 | 0.09 | 0.10 | 0.10 | 0.00 | 0.00 | 0.01 | 6.67 |
| **Australia** | 0.45 | 0.44 | 0.30 | 0.30 | 3.20 | 2.98 | 2.78 | 2.78 | 1.42 | 1.30 | 0.84 | 0.84 | 0.00 | 0.00 | -0.47 | -35.77 |
| **European Union - 28** | 0.36 | 0.31 | 0.35 | 0.35 | 1.36 | 1.54 | 1.55 | 1.55 | 0.48 | 0.48 | 0.54 | 0.54 | 0.00 | 0.00 | 0.06 | 12.53 |
| Greece | 0.29 | 0.25 | 0.27 | 0.27 | 1.33 | 1.63 | 1.71 | 1.71 | 0.38 | 0.40 | 0.45 | 0.45 | 0.00 | 0.00 | 0.05 | 13.25 |
| Spain | 0.07 | 0.06 | 0.08 | 0.08 | 1.47 | 1.22 | 1.06 | 1.06 | 0.10 | 0.08 | 0.09 | 0.09 | 0.00 | 0.00 | 0.01 | 8.97 |
| **Sub-Saharan Africa** | | | | | | | | | | | | | | | | |
| Burkina | 0.58 | 0.56 | 0.57 | 0.57 | 0.56 | 0.60 | 0.54 | 0.54 | 0.33 | 0.33 | 0.30 | 0.30 | 0.00 | 0.00 | -0.03 | -8.43 |
| Mali | 0.52 | 0.48 | 0.52 | 0.57 | 0.47 | 0.50 | 0.50 | 0.50 | 0.25 | 0.24 | 0.26 | 0.28 | 0.02 | 9.27 | 0.04 | 17.43 |
| Cameroon | 0.21 | 0.22 | 0.22 | 0.22 | 1.10 | 1.14 | 1.09 | 1.09 | 0.23 | 0.25 | 0.24 | 0.24 | 0.00 | 0.00 | -0.01 | -4.00 |
| Sudan | 0.06 | 0.06 | 0.06 | 0.06 | 1.07 | 0.84 | 0.84 | 0.84 | 0.06 | 0.05 | 0.05 | 0.05 | 0.00 | 0.00 | 0.00 | 0.00 |
| Zimbabwe | 0.24 | 0.23 | 0.30 | 0.30 | 0.41 | 0.50 | 0.47 | 0.47 | 0.10 | 0.11 | 0.14 | 0.14 | 0.00 | 0.00 | 0.03 | 23.68 |
| Nigeria | 0.30 | 0.29 | 0.30 | 0.30 | 0.47 | 0.45 | 0.46 | 0.46 | 0.14 | 0.13 | 0.14 | 0.14 | 0.00 | 0.00 | 0.01 | 6.15 |
| Benin | 0.25 | 0.27 | 0.27 | 0.27 | 0.73 | 0.61 | 0.61 | 0.61 | 0.18 | 0.17 | 0.17 | 0.17 | 0.00 | 0.00 | 0.00 | 0.00 |
| Uganda | 0.08 | 0.05 | 0.07 | 0.07 | 1.16 | 1.54 | 1.28 | 1.28 | 0.09 | 0.08 | 0.08 | 0.08 | 0.00 | 0.00 | 0.00 | 3.75 |
| **Egypt** | 0.14 | 0.13 | 0.16 | 0.16 | 1.04 | 1.02 | 1.03 | 1.03 | 0.15 | 0.13 | 0.16 | 0.16 | 0.00 | 0.00 | 0.03 | 20.30 |
| **Mexico** | 0.16 | 0.12 | 0.16 | 0.18 | 2.27 | 2.64 | 2.34 | 2.33 | 0.35 | 0.31 | 0.37 | 0.41 | 0.03 | 9.09 | 0.09 | 29.94 |
| **Burma** | 0.30 | 0.30 | 0.30 | 0.30 | 0.63 | 0.63 | 0.63 | 0.63 | 0.19 | 0.19 | 0.19 | 0.19 | 0.00 | 0.00 | 0.00 | 0.00 |
| **Others** | 1.51 | 1.35 | 1.35 | 1.33 | 0.53 | 0.52 | 0.52 | 0.52 | 0.79 | 0.70 | 0.70 | 0.69 | -0.01 | -1.42 | -0.01 | -1.28 |

World and Selected Countries and Regions

## Table 13 Peanut Area, Yield, and Production

| Country / Region | Area (Million hectares) | | | | Yield (Metric tons per hectare) | | | | Production (Million metric tons) | | | | Change in Production | | | |
|---|---|---|---|---|---|---|---|---|---|---|---|---|---|---|---|---|
| | | | 2014/15 Proj. | | | | 2014/15 Proj. | | | | 2014/15 Proj. | | From last month | | From last year | |
| | 2012/13 | Prel. 2013/14 | Jul | Aug | 2012/13 | Prel. 2013/14 | Jul | Aug | 2012/13 | Prel. 2013/14 | Jul | Aug | MMT | Percent | MMT | Percent |
| **World** | 23.94 | 23.71 | 23.61 | 23.59 | 1.68 | 1.68 | 1.71 | 1.70 | 40.12 | 39.79 | 40.40 | 40.18 | -0.22 | -0.54 | 0.40 | 1.00 |
| **United States** | 0.65 | 0.42 | 0.52 | 0.52 | 4.73 | 4.49 | 4.48 | 4.44 | 3.07 | 1.89 | 2.32 | 2.30 | -0.02 | -0.86 | 0.41 | 21.61 |
| **Total Foreign** | 23.29 | 23.29 | 23.09 | 23.07 | 1.59 | 1.63 | 1.65 | 1.64 | 37.06 | 37.90 | 38.08 | 37.88 | -0.20 | -0.53 | -0.01 | -0.03 |
| **China** | 4.64 | 4.71 | 4.72 | 4.70 | 3.60 | 3.60 | 3.64 | 3.62 | 16.69 | 16.97 | 17.20 | 17.00 | -0.20 | -1.16 | 0.03 | 0.16 |
| **South Asia** | | | | | | | | | | | | | | | | |
| India | 5.00 | 5.40 | 5.20 | 5.20 | 1.00 | 1.05 | 1.05 | 1.05 | 5.00 | 5.65 | 5.45 | 5.45 | 0.00 | 0.00 | -0.20 | -3.54 |
| Pakistan | 0.11 | 0.11 | 0.11 | 0.11 | 0.90 | 0.90 | 0.90 | 0.90 | 0.10 | 0.10 | 0.10 | 0.10 | 0.00 | 0.00 | 0.00 | 0.00 |
| **Sub-Saharan Africa** | 10.96 | 10.52 | 10.53 | 10.53 | 0.92 | 0.94 | 0.94 | 0.94 | 10.09 | 9.91 | 9.92 | 9.92 | 0.00 | 0.00 | 0.00 | 0.04 |
| Nigeria | 2.42 | 2.50 | 2.50 | 2.50 | 1.27 | 1.20 | 1.20 | 1.20 | 3.07 | 3.00 | 3.00 | 3.00 | 0.00 | 0.00 | 0.00 | 0.00 |
| Senegal | 0.71 | 0.77 | 0.77 | 0.77 | 0.98 | 0.92 | 0.94 | 0.94 | 0.69 | 0.71 | 0.73 | 0.73 | 0.00 | 0.00 | 0.02 | 2.11 |
| Chad | 0.41 | 0.50 | 0.50 | 0.50 | 0.90 | 0.80 | 0.80 | 0.80 | 0.37 | 0.40 | 0.40 | 0.40 | 0.00 | 0.00 | 0.00 | 0.00 |
| Ghana | 0.35 | 0.40 | 0.40 | 0.40 | 1.38 | 1.10 | 1.10 | 1.10 | 0.48 | 0.44 | 0.44 | 0.44 | 0.00 | 0.00 | 0.00 | 0.00 |
| Sudan | 1.62 | 1.00 | 1.00 | 1.00 | 0.64 | 0.85 | 0.85 | 0.85 | 1.03 | 0.85 | 0.85 | 0.85 | 0.00 | 0.00 | 0.00 | 0.00 |
| Congo (Kinshasa) | 0.48 | 0.48 | 0.48 | 0.48 | 0.78 | 0.78 | 0.78 | 0.78 | 0.37 | 0.37 | 0.37 | 0.37 | 0.00 | 0.00 | 0.00 | 0.00 |
| Burkina | 0.36 | 0.35 | 0.35 | 0.35 | 0.73 | 0.77 | 0.77 | 0.77 | 0.26 | 0.27 | 0.27 | 0.27 | 0.00 | 0.00 | 0.00 | 0.00 |
| Guinea | 0.22 | 0.21 | 0.21 | 0.21 | 1.38 | 1.24 | 1.24 | 1.24 | 0.30 | 0.26 | 0.26 | 0.26 | 0.00 | 0.00 | 0.00 | 0.00 |
| Cameroon | 0.41 | 0.40 | 0.40 | 0.40 | 1.39 | 1.38 | 1.38 | 1.38 | 0.57 | 0.55 | 0.55 | 0.55 | 0.00 | 0.00 | 0.00 | 0.00 |
| Mali | 0.34 | 0.35 | 0.35 | 0.35 | 0.95 | 0.93 | 0.93 | 0.93 | 0.33 | 0.33 | 0.33 | 0.33 | 0.00 | 0.00 | 0.00 | 0.00 |
| Malawi | 0.35 | 0.36 | 0.37 | 0.37 | 1.09 | 1.05 | 1.03 | 1.03 | 0.39 | 0.38 | 0.38 | 0.38 | 0.00 | 0.00 | -0.01 | -1.57 |
| Cote d'Ivoire | 0.08 | 0.08 | 0.08 | 0.08 | 1.21 | 1.13 | 1.13 | 1.13 | 0.09 | 0.09 | 0.09 | 0.09 | 0.00 | 0.00 | 0.00 | 0.00 |
| Uganda | 0.30 | 0.30 | 0.30 | 0.30 | 1.00 | 1.00 | 1.00 | 1.00 | 0.30 | 0.30 | 0.30 | 0.30 | 0.00 | 0.00 | 0.00 | 0.00 |
| Central African Republic | 0.10 | 0.10 | 0.10 | 0.10 | 1.55 | 1.50 | 1.50 | 1.50 | 0.15 | 0.15 | 0.15 | 0.15 | 0.00 | 0.00 | 0.00 | 0.00 |
| Benin | 0.13 | 0.13 | 0.13 | 0.13 | 0.65 | 0.65 | 0.65 | 0.65 | 0.08 | 0.09 | 0.09 | 0.09 | 0.00 | 0.00 | 0.00 | 0.00 |
| Mozambique | 0.39 | 0.29 | 0.29 | 0.29 | 0.29 | 0.38 | 0.38 | 0.38 | 0.11 | 0.11 | 0.11 | 0.11 | 0.00 | 0.00 | 0.00 | 0.00 |
| Niger | 0.74 | 0.74 | 0.74 | 0.74 | 0.39 | 0.49 | 0.47 | 0.47 | 0.29 | 0.37 | 0.35 | 0.35 | 0.00 | 0.00 | -0.02 | -4.11 |
| South Africa | 0.05 | 0.05 | 0.06 | 0.06 | 1.19 | 2.12 | 2.00 | 2.00 | 0.06 | 0.11 | 0.12 | 0.12 | 0.00 | 0.00 | 0.01 | 9.09 |
| **Southeast Asia** | | | | | | | | | | | | | | | | |
| Indonesia | 0.68 | 0.66 | 0.63 | 0.63 | 1.68 | 1.77 | 1.83 | 1.83 | 1.15 | 1.16 | 1.15 | 1.15 | 0.00 | 0.00 | -0.01 | -0.86 |
| Burma | 0.88 | 0.89 | 0.89 | 0.89 | 1.56 | 1.58 | 1.58 | 1.58 | 1.37 | 1.40 | 1.40 | 1.40 | 0.00 | 0.00 | 0.00 | 0.00 |
| Vietnam | 0.22 | 0.23 | 0.24 | 0.24 | 2.28 | 2.30 | 2.29 | 2.29 | 0.49 | 0.53 | 0.55 | 0.55 | 0.00 | 0.00 | 0.02 | 3.77 |
| Thailand | 0.03 | 0.03 | 0.03 | 0.03 | 1.53 | 1.67 | 1.67 | 1.67 | 0.05 | 0.05 | 0.05 | 0.05 | 0.00 | 0.00 | 0.00 | 0.00 |
| **South America** | | | | | | | | | | | | | | | | |
| Argentina | 0.38 | 0.33 | 0.34 | 0.34 | 2.67 | 2.94 | 3.53 | 3.53 | 1.02 | 0.98 | 1.20 | 1.20 | 0.00 | 0.00 | 0.22 | 22.20 |
| Brazil | 0.10 | 0.11 | 0.10 | 0.10 | 3.36 | 3.37 | 2.90 | 2.90 | 0.33 | 0.36 | 0.29 | 0.29 | 0.00 | 0.00 | -0.07 | -20.33 |
| **Egypt** | 0.06 | 0.06 | 0.06 | 0.06 | 3.17 | 3.17 | 3.17 | 3.17 | 0.19 | 0.19 | 0.19 | 0.19 | 0.00 | 0.00 | 0.00 | 0.00 |
| **Mexico** | 0.06 | 0.06 | 0.06 | 0.06 | 1.98 | 1.75 | 1.75 | 1.75 | 0.12 | 0.10 | 0.10 | 0.10 | 0.00 | 0.00 | 0.00 | 0.00 |
| **Others** | 0.18 | 0.20 | 0.20 | 0.20 | 2.58 | 2.50 | 2.51 | 2.51 | 0.47 | 0.49 | 0.49 | 0.49 | 0.00 | 0.00 | 0.00 | 0.20 |

World and Selected Countries and Regions

Table 14 Sunflowerseed Area, Yield, and Production

| Country / Region | Area (Million hectares) | | | | Yield (Metric tons per hectare) | | | | Production (Million metric tons) | | | | Change in Production | | | |
|---|---|---|---|---|---|---|---|---|---|---|---|---|---|---|---|---|
| | | | 2014/15 Proj. | | | | 2014/15 Proj. | | | | 2014/15 Proj. | | From last month | | From last year | |
| | 2012/13 | Prel. 2013/14 | Jul | Aug | 2012/13 | Prel. 2013/14 | Jul | Aug | 2012/13 | Prel. 2013/14 | Jul | Aug | MMT | Percent | MMT | Percent |
| World | 23.77 | 24.64 | 24.71 | 24.41 | 1.51 | 1.74 | 1.65 | 1.65 | 35.98 | 42.86 | 40.67 | 40.34 | -0.33 | -0.80 | -2.52 | -5.88 |
| United States | 0.75 | 0.60 | 0.66 | 0.66 | 1.70 | 1.54 | 1.66 | 1.66 | 1.26 | 0.92 | 1.10 | 1.10 | 0.00 | 0.00 | 0.17 | 18.87 |
| Total Foreign | 23.03 | 24.05 | 24.05 | 23.75 | 1.51 | 1.74 | 1.65 | 1.65 | 34.72 | 41.94 | 39.57 | 39.25 | -0.33 | -0.82 | -2.69 | -6.42 |
| Former Soviet Union - 12 | 12.61 | 13.52 | 13.60 | 13.30 | 1.40 | 1.72 | 1.56 | 1.57 | 17.66 | 23.23 | 21.25 | 20.85 | -0.40 | -1.88 | -2.38 | -10.23 |
| Russia | 6.13 | 6.80 | 6.80 | 6.50 | 1.30 | 1.55 | 1.50 | 1.51 | 7.96 | 10.55 | 10.20 | 9.80 | -0.40 | -3.92 | -0.75 | -7.14 |
| Ukraine | 5.50 | 5.60 | 5.60 | 5.60 | 1.64 | 2.07 | 1.79 | 1.79 | 9.00 | 11.60 | 10.00 | 10.00 | 0.00 | 0.00 | -1.60 | -13.79 |
| Moldova | 0.30 | 0.30 | 0.30 | 0.30 | 0.99 | 1.67 | 1.50 | 1.50 | 0.30 | 0.50 | 0.45 | 0.45 | 0.00 | 0.00 | -0.05 | -9.82 |
| Kazakhstan | 0.68 | 0.82 | 0.90 | 0.90 | 0.59 | 0.70 | 0.67 | 0.67 | 0.40 | 0.57 | 0.60 | 0.60 | 0.00 | 0.00 | 0.03 | 4.71 |
| South America | 1.94 | 1.69 | 2.02 | 2.02 | 1.85 | 1.59 | 1.76 | 1.76 | 3.60 | 2.68 | 3.55 | 3.55 | 0.00 | 0.00 | 0.87 | 32.43 |
| Argentina | 1.62 | 1.30 | 1.60 | 1.60 | 1.91 | 1.62 | 1.81 | 1.81 | 3.10 | 2.10 | 2.90 | 2.90 | 0.00 | 0.00 | 0.80 | 38.10 |
| Uruguay | 0.03 | 0.03 | 0.03 | 0.03 | 1.60 | 1.40 | 1.47 | 1.47 | 0.04 | 0.04 | 0.04 | 0.04 | 0.00 | 0.00 | 0.01 | 25.71 |
| Bolivia | 0.17 | 0.18 | 0.20 | 0.20 | 1.47 | 1.50 | 1.50 | 1.50 | 0.25 | 0.27 | 0.30 | 0.30 | 0.00 | 0.00 | 0.03 | 11.11 |
| Brazil | 0.07 | 0.13 | 0.13 | 0.13 | 1.57 | 1.50 | 1.54 | 1.54 | 0.11 | 0.20 | 0.20 | 0.20 | 0.00 | 0.00 | 0.01 | 2.56 |
| Paraguay | 0.05 | 0.05 | 0.06 | 0.06 | 1.78 | 1.60 | 1.69 | 1.69 | 0.10 | 0.08 | 0.11 | 0.11 | 0.00 | 0.00 | 0.03 | 31.25 |
| China | 0.89 | 0.97 | 0.98 | 0.98 | 2.61 | 2.54 | 2.56 | 2.56 | 2.32 | 2.45 | 2.50 | 2.50 | 0.00 | 0.00 | 0.05 | 2.04 |
| European Union - 28 | 4.29 | 4.51 | 4.20 | 4.20 | 1.65 | 1.97 | 1.89 | 1.89 | 7.07 | 8.88 | 7.93 | 7.95 | 0.03 | 0.32 | -0.93 | -10.43 |
| France | 0.68 | 0.77 | 0.72 | 0.68 | 2.31 | 2.05 | 2.33 | 2.37 | 1.57 | 1.58 | 1.68 | 1.60 | -0.08 | -4.48 | 0.02 | 1.27 |
| Hungary | 0.61 | 0.59 | 0.60 | 0.60 | 2.14 | 2.47 | 2.29 | 2.25 | 1.30 | 1.47 | 1.38 | 1.35 | -0.03 | -1.82 | -0.12 | -8.16 |
| Spain | 0.75 | 0.85 | 0.80 | 0.82 | 0.82 | 1.21 | 1.00 | 1.10 | 0.62 | 1.03 | 0.80 | 0.91 | 0.11 | 13.75 | -0.12 | -11.56 |
| Italy | 0.11 | 0.11 | 0.11 | 0.11 | 1.71 | 2.25 | 2.13 | 2.19 | 0.19 | 0.24 | 0.23 | 0.24 | 0.01 | 3.04 | -0.01 | -2.47 |
| Slovakia | 0.09 | 0.08 | 0.09 | 0.08 | 2.19 | 2.28 | 2.27 | 2.30 | 0.20 | 0.19 | 0.20 | 0.19 | -0.01 | -4.62 | 0.00 | -1.59 |
| South Asia | 1.27 | 1.15 | 1.09 | 1.09 | 1.10 | 1.10 | 1.09 | 1.09 | 1.40 | 1.27 | 1.19 | 1.19 | 0.00 | 0.00 | -0.08 | -6.30 |
| India | 0.80 | 0.75 | 0.73 | 0.73 | 0.88 | 0.89 | 0.89 | 0.89 | 0.70 | 0.67 | 0.65 | 0.65 | 0.00 | 0.00 | -0.02 | -2.99 |
| Pakistan | 0.47 | 0.40 | 0.36 | 0.36 | 1.49 | 1.50 | 1.50 | 1.50 | 0.70 | 0.60 | 0.54 | 0.54 | 0.00 | 0.00 | -0.06 | -10.00 |
| Turkey | 0.60 | 0.69 | 0.60 | 0.60 | 1.88 | 2.03 | 1.92 | 1.92 | 1.13 | 1.40 | 1.15 | 1.15 | 0.00 | 0.00 | -0.25 | -17.86 |
| South Africa | 0.51 | 0.60 | 0.61 | 0.61 | 1.10 | 1.42 | 1.31 | 1.31 | 0.56 | 0.85 | 0.80 | 0.80 | 0.00 | 0.00 | -0.05 | -6.21 |
| Burma | 0.54 | 0.54 | 0.54 | 0.54 | 0.65 | 0.90 | 0.90 | 0.90 | 0.35 | 0.49 | 0.49 | 0.49 | 0.00 | 0.00 | 0.00 | 0.00 |
| Serbia | 0.18 | 0.19 | 0.19 | 0.19 | 2.00 | 2.30 | 2.30 | 2.57 | 0.35 | 0.43 | 0.43 | 0.48 | 0.05 | 11.76 | 0.05 | 11.76 |
| Canada | 0.04 | 0.03 | 0.03 | 0.03 | 2.18 | 1.86 | 1.76 | 1.76 | 0.09 | 0.05 | 0.06 | 0.06 | 0.00 | 0.00 | 0.01 | 15.38 |
| Australia | 0.03 | 0.03 | 0.04 | 0.04 | 1.47 | 1.19 | 1.28 | 1.28 | 0.04 | 0.03 | 0.05 | 0.05 | 0.00 | 0.00 | 0.01 | 43.75 |
| Others | 0.14 | 0.16 | 0.16 | 0.16 | 1.17 | 1.23 | 1.23 | 1.23 | 0.17 | 0.19 | 0.19 | 0.19 | 0.00 | 0.00 | 0.00 | 0.00 |

World and Selected Countries and Regions

# Table 15 Rapeseed Area, Yield, and Production

| Country / Region | Area (Million hectares) | | | | Yield (Metric tons per hectare) | | | | Production (Million metric tons) | | | | Change in Production | | | |
|---|---|---|---|---|---|---|---|---|---|---|---|---|---|---|---|---|
| | | | 2014/15 Proj. | | | | 2014/15 Proj. | | | | 2014/15 Proj. | | From last month | | From last year | |
| | 2012/13 | Prel. 2013/14 | Jul | Aug | 2012/13 | Prel. 2013/14 | Jul | Aug | 2012/13 | Prel. 2013/14 | Jul | Aug | MMT | Percent | MMT | Percent |
| **World** | 36.30 | 36.20 | 36.68 | 36.21 | 1.75 | 1.97 | 1.91 | 1.94 | 63.67 | 71.20 | 70.16 | 70.36 | 0.20 | 0.29 | -0.84 | -1.19 |
| **United States** | 0.70 | 0.51 | 0.68 | 0.68 | 1.59 | 1.96 | 1.76 | 1.76 | 1.11 | 1.00 | 1.19 | 1.19 | 0.00 | 0.00 | 0.19 | 18.73 |
| **Total Foreign** | 35.59 | 35.68 | 36.00 | 35.53 | 1.76 | 1.97 | 1.92 | 1.95 | 62.56 | 70.20 | 68.97 | 69.17 | 0.20 | 0.29 | -1.03 | -1.47 |
| **European Union - 28** | 6.33 | 6.75 | 6.88 | 6.81 | 3.10 | 3.12 | 3.26 | 3.33 | 19.63 | 21.10 | 22.40 | 22.65 | 0.25 | 1.12 | 1.55 | 7.34 |
| Germany | 1.31 | 1.46 | 1.44 | 1.44 | 3.69 | 3.95 | 4.10 | 4.10 | 4.82 | 5.78 | 5.88 | 5.88 | 0.00 | 0.00 | 0.10 | 1.77 |
| France | 1.61 | 1.44 | 1.52 | 1.50 | 3.40 | 3.04 | 3.44 | 3.53 | 5.46 | 4.37 | 5.23 | 5.30 | 0.08 | 1.44 | 0.93 | 21.28 |
| United Kingdom | 0.76 | 0.72 | 0.71 | 0.72 | 3.38 | 2.97 | 3.58 | 3.57 | 2.56 | 2.13 | 2.53 | 2.57 | 0.04 | 1.62 | 0.44 | 20.83 |
| Poland | 0.84 | 0.96 | 0.90 | 0.86 | 2.65 | 2.92 | 2.98 | 3.09 | 2.22 | 2.80 | 2.68 | 2.66 | -0.02 | -0.75 | -0.14 | -5.00 |
| Czech Republic | 0.40 | 0.42 | 0.39 | 0.39 | 2.77 | 3.45 | 3.16 | 3.24 | 1.11 | 1.44 | 1.24 | 1.26 | 0.02 | 1.53 | -0.18 | -12.75 |
| Denmark | 0.13 | 0.18 | 0.18 | 0.19 | 3.71 | 3.85 | 3.89 | 3.80 | 0.48 | 0.68 | 0.70 | 0.70 | 0.00 | 0.43 | 0.02 | 3.23 |
| Hungary | 0.17 | 0.20 | 0.23 | 0.23 | 2.40 | 2.59 | 2.43 | 2.48 | 0.40 | 0.52 | 0.56 | 0.57 | 0.01 | 1.79 | 0.05 | 8.78 |
| Romania | 0.11 | 0.29 | 0.42 | 0.43 | 1.53 | 2.45 | 2.26 | 2.37 | 0.17 | 0.71 | 0.95 | 1.02 | 0.07 | 7.37 | 0.31 | 43.66 |
| Slovakia | 0.11 | 0.14 | 0.12 | 0.13 | 1.99 | 2.73 | 2.54 | 2.98 | 0.21 | 0.37 | 0.31 | 0.37 | 0.06 | 20.32 | 0.00 | -0.27 |
| Sweden | 0.11 | 0.12 | 0.13 | 0.10 | 2.97 | 2.73 | 2.72 | 3.14 | 0.32 | 0.33 | 0.34 | 0.30 | -0.04 | -12.35 | -0.04 | -10.51 |
| Lithuania | 0.26 | 0.26 | 0.26 | 0.25 | 2.43 | 2.13 | 2.31 | 2.22 | 0.63 | 0.55 | 0.60 | 0.55 | -0.05 | -8.00 | 0.00 | 0.36 |
| Latvia | 0.12 | 0.13 | 0.12 | 0.11 | 2.58 | 2.32 | 2.30 | 2.32 | 0.30 | 0.30 | 0.27 | 0.26 | -0.01 | -3.77 | -0.04 | -14.14 |
| Austria | 0.06 | 0.06 | 0.06 | 0.05 | 2.64 | 3.40 | 3.10 | 3.49 | 0.15 | 0.20 | 0.18 | 0.19 | 0.01 | 2.78 | -0.01 | -6.09 |
| Finland | 0.06 | 0.05 | 0.06 | 0.06 | 1.25 | 1.60 | 1.36 | 1.64 | 0.08 | 0.09 | 0.08 | 0.09 | 0.02 | 20.00 | 0.01 | 5.88 |
| Estonia | 0.09 | 0.09 | 0.09 | 0.09 | 1.82 | 2.02 | 1.98 | 1.98 | 0.16 | 0.17 | 0.17 | 0.17 | 0.00 | -1.16 | 0.00 | -2.30 |
| **China** | 7.43 | 7.52 | 7.50 | 7.50 | 1.88 | 1.92 | 1.93 | 1.96 | 14.01 | 14.46 | 14.50 | 14.70 | 0.20 | 1.38 | 0.24 | 1.67 |
| **South Asia** | | | | | | | | | | | | | | | | |
| India | 6.75 | 7.13 | 7.20 | 7.20 | 1.01 | 1.02 | 1.04 | 1.04 | 6.80 | 7.30 | 7.50 | 7.50 | 0.00 | 0.00 | 0.20 | 2.74 |
| Pakistan | 0.38 | 0.36 | 0.35 | 0.35 | 0.92 | 0.89 | 0.86 | 0.86 | 0.35 | 0.32 | 0.30 | 0.30 | 0.00 | 0.00 | -0.02 | -6.25 |
| Bangladesh | 0.31 | 0.31 | 0.31 | 0.31 | 0.74 | 0.74 | 0.74 | 0.74 | 0.23 | 0.23 | 0.23 | 0.23 | 0.00 | 0.00 | 0.00 | 0.00 |
| **Canada** | 8.80 | 8.01 | 8.00 | 7.70 | 1.58 | 2.25 | 1.96 | 1.98 | 13.87 | 18.00 | 15.70 | 15.25 | -0.45 | -2.87 | -2.75 | -15.28 |
| **Australia** | 3.27 | 2.66 | 2.70 | 2.70 | 1.27 | 1.42 | 1.28 | 1.28 | 4.14 | 3.76 | 3.45 | 3.45 | 0.00 | 0.00 | -0.31 | -8.24 |
| **Former Soviet Union - 12** | | | | | | | | | | | | | | | | |
| Ukraine | 0.55 | 1.00 | 0.95 | 0.85 | 2.38 | 2.36 | 2.11 | 2.59 | 1.30 | 2.35 | 2.00 | 2.20 | 0.20 | 10.00 | -0.15 | -6.46 |
| Russia | 0.97 | 1.11 | 1.20 | 1.20 | 1.07 | 1.25 | 1.25 | 1.25 | 1.04 | 1.39 | 1.50 | 1.50 | 0.00 | 0.00 | 0.11 | 7.68 |
| Belarus | 0.42 | 0.40 | 0.43 | 0.43 | 1.67 | 1.68 | 1.65 | 1.65 | 0.71 | 0.68 | 0.70 | 0.70 | 0.00 | 0.00 | 0.02 | 3.55 |
| Paraguay | 0.08 | 0.09 | 0.09 | 0.09 | 1.37 | 1.41 | 1.40 | 1.40 | 0.12 | 0.12 | 0.13 | 0.13 | 0.00 | 0.00 | 0.01 | 5.00 |
| Others | 0.30 | 0.35 | 0.40 | 0.40 | 1.25 | 1.41 | 1.41 | 1.41 | 0.37 | 0.49 | 0.56 | 0.56 | 0.00 | 0.00 | 0.07 | 14.96 |

World and Selected Countries and Regions

# Table 16 Copra, Palm Kernel, and Palm Oil Production

| Country / Region | Production (Million metric tons) | | | | Change in Production | | | |
| --- | --- | --- | --- | --- | --- | --- | --- | --- |
| | 2012/13 | Prel. 2013/14 | 2014/15 Proj. Jul | 2014/15 Proj. Aug | From last month MMT | From last month Percent | From last year MMT | From last year Percent |
| **Oilseed, Copra** | | | | | | | | |
| Philippines | 2.65 | 2.40 | 2.35 | 2.35 | 0.00 | 0.00 | -0.05 | -2.08 |
| Indonesia | 1.56 | 1.58 | 1.58 | 1.58 | 0.00 | 0.00 | 0.00 | 0.00 |
| India | 0.67 | 0.67 | 0.67 | 0.67 | 0.00 | 0.00 | 0.00 | 0.00 |
| Vietnam | 0.24 | 0.24 | 0.24 | 0.24 | 0.00 | 0.00 | 0.00 | 0.00 |
| Mexico | 0.21 | 0.21 | 0.21 | 0.21 | 0.00 | 0.00 | 0.00 | 0.00 |
| Papua New Guinea | 0.13 | 0.13 | 0.13 | 0.13 | 0.00 | 0.00 | 0.00 | 0.00 |
| Thailand | 0.07 | 0.07 | 0.07 | 0.07 | 0.00 | 0.00 | 0.00 | 0.00 |
| Sri Lanka | 0.07 | 0.07 | 0.07 | 0.07 | 0.00 | 0.00 | 0.00 | 0.00 |
| Solomon Islands | 0.03 | 0.03 | 0.03 | 0.03 | 0.00 | 0.00 | 0.00 | 0.00 |
| Cote d'Ivoire | 0.03 | 0.03 | 0.03 | 0.03 | 0.00 | 0.00 | 0.00 | 0.00 |
| World | 5.80 | 5.58 | 5.53 | 5.53 | 0.00 | 0.00 | -0.05 | -0.90 |
| | | | | | | | | |
| **Oilseed, Palm Kernel** | | | | | | | | |
| Indonesia | 7.48 | 8.14 | 8.70 | 8.70 | 0.00 | 0.00 | 0.56 | 6.88 |
| Malaysia | 4.87 | 5.00 | 5.20 | 5.20 | 0.00 | 0.00 | 0.20 | 4.00 |
| Nigeria | 0.70 | 0.73 | 0.73 | 0.73 | 0.00 | 0.00 | 0.01 | 1.39 |
| Thailand | 0.43 | 0.44 | 0.45 | 0.45 | 0.00 | 0.00 | 0.01 | 2.27 |
| Colombia | 0.21 | 0.24 | 0.24 | 0.25 | 0.01 | 4.17 | 0.01 | 4.17 |
| Papua New Guinea | 0.13 | 0.13 | 0.13 | 0.13 | 0.00 | 0.00 | 0.00 | 0.00 |
| Cameroon | 0.11 | 0.11 | 0.11 | 0.11 | 0.00 | 0.00 | 0.00 | 0.00 |
| Ecuador | 0.11 | 0.11 | 0.12 | 0.12 | 0.00 | 0.00 | 0.01 | 9.09 |
| Honduras | 0.10 | 0.10 | 0.11 | 0.11 | 0.00 | 0.00 | 0.01 | 10.00 |
| Brazil | 0.10 | 0.10 | 0.10 | 0.10 | 0.00 | 0.00 | 0.00 | 0.00 |
| World | 14.79 | 15.66 | 16.46 | 16.47 | 0.01 | 0.06 | 0.81 | 5.17 |
| | | | | | | | | |
| **Oil, Palm** | | | | | | | | |
| Indonesia | 28.50 | 31.00 | 33.50 | 33.50 | 0.00 | 0.00 | 2.50 | 8.06 |
| Malaysia | 19.32 | 19.90 | 20.80 | 20.80 | 0.00 | 0.00 | 0.90 | 4.52 |
| Thailand | 2.14 | 2.15 | 2.25 | 2.25 | 0.00 | 0.00 | 0.10 | 4.65 |
| Colombia | 0.97 | 1.04 | 1.03 | 1.07 | 0.05 | 4.90 | 0.03 | 2.88 |
| Nigeria | 0.91 | 0.93 | 0.93 | 0.93 | 0.00 | 0.00 | 0.00 | 0.00 |
| Papua New Guinea | 0.61 | 0.63 | 0.63 | 0.63 | 0.00 | 0.00 | 0.00 | 0.00 |
| Ecuador | 0.54 | 0.57 | 0.58 | 0.58 | 0.00 | 0.00 | 0.01 | 1.79 |
| Honduras | 0.41 | 0.43 | 0.44 | 0.44 | 0.00 | 0.00 | 0.01 | 2.33 |
| Cote d'Ivoire | 0.39 | 0.40 | 0.40 | 0.40 | 0.00 | 0.00 | 0.00 | 0.00 |
| Guatemala | 0.32 | 0.35 | 0.36 | 0.36 | 0.00 | 0.00 | 0.01 | 2.86 |
| World | 55.97 | 59.30 | 62.80 | 62.84 | 0.04 | 0.06 | 3.54 | 5.97 |

World and Selected Countries and Regions

## Table 17 Cotton Area, Yield, and Production

| Country / Region | Area (Million hectares) | | | | Yield (Kilograms per hectare) | | | | Production (Million 480 lb. bales) | | | | Change in Production | | | |
|---|---|---|---|---|---|---|---|---|---|---|---|---|---|---|---|---|
| | 2012/13 | Prel. 2013/14 | 2014/15 Proj. Jul | 2014/15 Proj. Aug | 2012/13 | Prel. 2013/14 | 2014/15 Proj. Jul | 2014/15 Proj. Aug | 2012/13 | Prel. 2013/14 | 2014/15 Proj. Jul | 2014/15 Proj. Aug | From last month MBales | Percent | From last year MBales | Percent |
| **World** | 34.33 | 32.77 | 33.39 | 33.93 | 780 | 786 | 759 | 755 | 122.95 | 118.27 | 116.42 | 117.64 | 1.22 | 1.05 | -0.63 | -0.53 |
| **United States** | 3.79 | 3.05 | 3.93 | 4.14 | 994 | 921 | 915 | 920 | 17.32 | 12.91 | 16.50 | 17.50 | 1.00 | 6.07 | 4.59 | 35.58 |
| **Total Foreign** | 30.54 | 29.71 | 29.46 | 29.79 | 753 | 772 | 738 | 732 | 105.64 | 105.36 | 99.92 | 100.14 | 0.22 | 0.22 | -5.23 | -4.96 |
| **South Asia** | | | | | | | | | | | | | | | | |
| India | 12.00 | 11.70 | 11.80 | 12.20 | 517 | 568 | 517 | 518 | 28.50 | 30.50 | 28.00 | 29.00 | 1.00 | 3.57 | -1.50 | -4.92 |
| Pakistan | 3.00 | 3.00 | 3.00 | 3.00 | 675 | 689 | 689 | 689 | 9.30 | 9.50 | 9.50 | 9.50 | 0.00 | 0.00 | 0.00 | 0.00 |
| **Former Soviet Union - 12** | | | | | | | | | | | | | | | | |
| Uzbekistan | 1.32 | 1.29 | 1.29 | 1.29 | 745 | 703 | 712 | 712 | 4.50 | 4.15 | 4.20 | 4.20 | 0.00 | 0.00 | 0.05 | 1.20 |
| Turkmenistan | 0.60 | 0.58 | 0.55 | 0.55 | 581 | 568 | 579 | 579 | 1.60 | 1.50 | 1.45 | 1.45 | 0.00 | 0.00 | -0.05 | -3.33 |
| Tajikistan | 0.20 | 0.19 | 0.19 | 0.18 | 599 | 516 | 588 | 538 | 0.55 | 0.45 | 0.50 | 0.45 | -0.05 | -10.00 | 0.00 | 0.00 |
| Kazakhstan | 0.15 | 0.14 | 0.13 | 0.13 | 611 | 536 | 544 | 544 | 0.42 | 0.34 | 0.33 | 0.33 | 0.00 | 0.00 | -0.02 | -4.41 |
| **Sub-Saharan Africa** | | | | | | | | | | | | | | | | |
| Burkina | 0.58 | 0.56 | 0.57 | 0.57 | 441 | 471 | 424 | 424 | 1.18 | 1.20 | 1.10 | 1.10 | 0.00 | 0.00 | -0.10 | -8.33 |
| Mali | 0.52 | 0.48 | 0.52 | 0.57 | 362 | 386 | 383 | 382 | 0.87 | 0.85 | 0.92 | 1.00 | 0.09 | 9.29 | 0.15 | 17.65 |
| Zimbabwe | 0.24 | 0.23 | 0.30 | 0.30 | 236 | 284 | 269 | 269 | 0.26 | 0.30 | 0.37 | 0.37 | 0.00 | 0.00 | 0.07 | 23.33 |
| Benin | 0.25 | 0.27 | 0.27 | 0.27 | 479 | 403 | 403 | 403 | 0.55 | 0.50 | 0.50 | 0.50 | 0.00 | 0.00 | 0.00 | 0.00 |
| Cote d'Ivoire | 0.38 | 0.36 | 0.37 | 0.37 | 398 | 460 | 403 | 403 | 0.69 | 0.75 | 0.68 | 0.68 | 0.00 | 0.00 | -0.08 | -10.00 |
| Cameroon | 0.21 | 0.22 | 0.22 | 0.22 | 477 | 495 | 475 | 475 | 0.46 | 0.50 | 0.48 | 0.48 | 0.00 | 0.00 | -0.02 | -4.00 |
| Nigeria | 0.30 | 0.29 | 0.30 | 0.30 | 236 | 225 | 232 | 232 | 0.33 | 0.30 | 0.32 | 0.32 | 0.00 | 0.00 | 0.02 | 6.67 |
| Sudan | 0.06 | 0.06 | 0.06 | 0.06 | 454 | 356 | 356 | 356 | 0.13 | 0.09 | 0.09 | 0.09 | 0.00 | 0.00 | 0.00 | 0.00 |
| **South America** | | | | | | | | | | | | | | | | |
| Brazil | 0.90 | 1.12 | 1.15 | 1.05 | 1,452 | 1,516 | 1,515 | 1,514 | 6.00 | 7.80 | 8.00 | 7.30 | -0.70 | -8.75 | -0.50 | -6.41 |
| Argentina | 0.36 | 0.56 | 0.57 | 0.57 | 454 | 474 | 472 | 472 | 0.75 | 1.22 | 1.23 | 1.23 | 0.00 | 0.00 | 0.01 | 0.41 |
| Paraguay | 0.05 | 0.03 | 0.04 | 0.03 | 363 | 392 | 354 | 392 | 0.08 | 0.05 | 0.07 | 0.05 | -0.02 | -30.77 | 0.00 | 0.00 |
| **Middle East** | | | | | | | | | | | | | | | | |
| Turkey | 0.41 | 0.33 | 0.42 | 0.42 | 1,407 | 1,517 | 1,503 | 1,503 | 2.65 | 2.30 | 2.90 | 2.90 | 0.00 | 0.00 | 0.60 | 26.09 |
| Syria | 0.13 | 0.10 | 0.06 | 0.06 | 1,263 | 1,306 | 1,089 | 1,089 | 0.73 | 0.60 | 0.30 | 0.30 | 0.00 | 0.00 | -0.30 | -50.00 |
| Iran | 0.11 | 0.12 | 0.11 | 0.11 | 591 | 606 | 594 | 594 | 0.29 | 0.32 | 0.30 | 0.30 | 0.00 | 0.00 | -0.02 | -6.25 |
| **Australia** | 0.45 | 0.44 | 0.30 | 0.28 | 2,251 | 2,047 | 1,960 | 1,979 | 4.60 | 4.10 | 2.70 | 2.50 | -0.20 | -7.41 | -1.60 | -39.02 |
| **European Union - 28** | 0.36 | 0.31 | 0.34 | 0.34 | 909 | 1,119 | 1,079 | 1,079 | 1.49 | 1.60 | 1.69 | 1.69 | 0.00 | 0.00 | 0.09 | 5.38 |
| Greece | 0.29 | 0.25 | 0.27 | 0.27 | 912 | 1,217 | 1,150 | 1,150 | 1.19 | 1.37 | 1.40 | 1.40 | 0.00 | 0.00 | 0.03 | 2.26 |
| Spain | 0.07 | 0.06 | 0.07 | 0.07 | 893 | 748 | 820 | 820 | 0.29 | 0.22 | 0.28 | 0.28 | 0.00 | 0.00 | 0.06 | 25.00 |
| **Egypt** | 0.14 | 0.13 | 0.16 | 0.16 | 746 | 729 | 737 | 737 | 0.49 | 0.44 | 0.53 | 0.53 | 0.00 | 0.00 | 0.09 | 20.69 |
| **Mexico** | 0.16 | 0.12 | 0.16 | 0.18 | 1,455 | 1,691 | 1,497 | 1,493 | 1.04 | 0.92 | 1.10 | 1.20 | 0.10 | 9.09 | 0.28 | 29.87 |
| **Others** | 7.69 | 7.13 | 6.62 | 6.62 | 1,082 | 1,072 | 1,075 | 1,075 | 38.22 | 35.09 | 32.70 | 32.70 | 0.00 | 0.00 | -2.39 | -6.82 |

World and Selected Countries and Regions

# TABLE 18

The table below presents a record of the differences between the August projection and the final Estimate. Using world wheat production as an example, the "root mean square error" means that chances are 2 out of 3 that the current forecast will not be above or below the final estimate by more than 2.4 percent. Chances are 9 out of 10 (90% confidence level) that the difference will not exceed 4.9 percent. The average difference between the August projection and the final estimate is 11.1 million tons, ranging from 0.2 million to 32.1 million tons. The August projection has been below the estimate 21 times and above 12 times.

## RELIABILITY OF PRODUCTION PROJECTIONS 1/

| COMMODITY AND REGION | Root mean square error | 90 percent confidence interval | Difference between forecast and final estimate | | | Years | |
|---|---|---|---|---|---|---|---|
| | | | Average | Smallest | Largest | Below final | Above final |
| | Percent | | ---Million metric tons--- | | | | |
| WHEAT | | | | | | | |
| World | 2.4 | 4.1 | 11.1 | 0 2 | 32.1 | 21 | 12 |
| U.S. | 2.4 | 4.0 | 1.1 | 0.0 | 4.2 | 14 | 19 |
| Foreign | 2.7 | 4.6 | 10.8 | 0.6 | 31.1 | 21 | 12 |
| COARSE GRAINS 2/ | | | | | | | |
| World | 2.4 | 4.1 | 17.1 | 0.4 | 51.0 | 25 | 8 |
| U.S. | 6.5 | 11.1 | 10.4 | 0.0 | 31.4 | 21 | 12 |
| Foreign | 2.4 | 4.0 | 12.7 | 0.7 | 28.9 | 23 | 10 |
| RICE (Milled) | | | | | | | |
| World | 2.4 | 4.1 | 6.7 | 0 1 | 24.4 | 22 | 11 |
| U.S. | 5.2 | 8.8 | 0.2 | 0.0 | 0.6 | 19 | 14 |
| Foreign | 2.5 | 4.2 | 6.7 | 0.4 | 24.7 | 23 | 10 |
| SOYBEANS | | | | | | | |
| World | 4.8 | 8.2 | 6.7 | 0 3 | 26.7 | 20 | 13 |
| U.S. | 6.4 | 10.9 | 3.4 | 0.0 | 11.1 | 17 | 15 |
| Foreign | 8.0 | 13.5 | 6.4 | 1 1 | 26.4 | 17 | 16 |
| COTTON | | | ---Million 480-lb. bales--- | | | | |
| World | 4.9 | 8.3 | 3.2 | 0.0 | 13.2 | 18 | 14 |
| U.S. | 7.8 | 13.3 | 1.0 | 0.0 | 3.9 | 17 | 15 |
| Foreign | 5.3 | 8.9 | 2.8 | 0.0 | 10.7 | 18 | 14 |
| UNITED STATES | | | -------Million bushels------- | | | | |
| CORN | 7.0 | 12..0 | 386 | 1 | 1,079 | 21 | 12 |
| SORGHUM | 8.5 | 14.5 | 36 | 1 | 108 | 17 | 16 |
| BARLEY | 6.3 | 10.8 | 16 | 1 | 67 | 11 | 22 |
| OATS | 9.3 | 15.7 | 14 | 1 | 57 | 6 | 27 |

1/ Marketing years 1981/82 through 2013/14. Final for grains, soybeans and cotton is defined as the first November estimates following the marketing year for 1981/82 through 2012/13, and for 2013/14 last month's estimate.

2/ Includes corn, sorghum, barley, oats, rye, millet, and mixed grain

August 2014                                                    Office of Global Analysis, FAS, USDA

www.ingramcontent.com/pod-product-compliance
Lightning Source LLC
Chambersburg PA
CBHW050431180526

45159CB00005B/2497